**DIMITRIOS E. BAIRAKTARIS**

# THE $8x2^n$ DIMENSIONS OF THE PARTICLES

| | |
|---|---|
| from | STRUCTURAL MECHANICS |
| to | PARTICLE PHYSICS |
| through | OCTONION MATHEMATICS |

LONDON 2021

DIMITRIOS E. BAIRAKTARIS

"THE $8x2^n$ DIMENSIONS OF THE PARTICLES"

ISBN: 978-1-912935-55-0



Cover photo: By Daniil Peshkov, copyright protected.
Editing & Pagination: Emmanuel D. Bairaktaris & Orianna Bairaktari



Printed by European Printers Ltd

Published by AKAKIA Publications

akakia

46 Warberry Road,
N22 7TQ, London, UK
T. 0044 208 2457 849
publications@akakia.net
www.akakia.net

# Table of Contents

## PREFACE

The author of this treatise is a structural engineer and former lecturer of structural mechanics in the Technical University of Athens, with fifty nine years of professional activity in the research on topics of the engineering practice and the design of a large variety of structures. At the same time he was employed as an amateur mathematician with a strong interest for applications, initially in theoretical mechanics, which was extended with time in general physics. Both occupations formed the basis for the present results, fruit of a fifty year effort.

There are many reasons for this long duration. First, the investigation was executed at the intervals among professional obligations. Secondly, the final choice for a career oriented to research and design, against the academic one, lessened the interest for frequent preparation of papers and publications, with priority given to the advancing of self-instruction in mathematics and physics and to the personal satisfaction for the professional achievements. But the most intense influence is attributed to the generative nature of this initiative. Generalizations of the initial ideas were continuously discovered, which imposed the revision of the previous items, accompanied by successive changes of the terminology, the notations and the symbolism, in order for a reasonable unification and simplification, adapted to the complexity of the overall created system, to be obtained.

Starting point of this undertaking was the suspicion of a relationship between the octonions and the finite set of the state parameters, which define the behaviour at any position along the members constituting the structural systems. For the vast majority of them, six state parameters are taken into account, consisting of three polar and three axial vector components, representing forces and moments respectively and the conjugated with them translations and rotations. For a specific type of structural member, for example beams with open hollow thin walled cross section, a seventh state parameter, referred as bi-moment, is taken into account. For reasons of simplification its influence on the other types of structural members is omitted, because of its negligible contribution to their total potential energy. After discovery of its vector character, not previously defined, which consists of a triple vector product equal to an imaginary scalar, the one to one correspondence between the known seven state parameters and the seven of the eight octonion components, after the exception of the real scalar one, became apparent. The revelation of an eight state parameter, represented by a real entity, was proven to be a relatively easy attempt and restored the complete one-to-one correspondence between the two eight-folds. Its total ignorance in the past is likewise justified from its negligible effect on the potential energy.

State parameters which, although active, are omitted at the treatment of some systems, because of the negligible effect they cause in the behaviour of those systems, possibly play a primitive role in the behaviour of other systems. The structural systems, which belong to the subject of the deformable continuum mechanics, offer the possibility for the recovery of the whole set of the parameters, which are required for the complete definition of their behaviour, without exceptions. A set of 32 parameters is the result if this investigation, which are distributed in four subsets, with each one containing eight members represented by octonion components. The reformation and extension of the octonions' algebra and analysis in a form suitable for the requirements of this innovation becomes inevitable. The detailed description of this formalism constitutes the content of this treatise. The application in specific systems, for example the internal structure of the elementary particles, considered to consist part of a deformable continuous substance, is the subject of an investigation procedure in progress aiming to a future presentation. The present treatise does not consist a contribution to topics of the science in phase of unfolded research, accompanied by publications or progress reports. It is based exclusively on the established knowledge supplied from the great number of the textbooks on mathematics, structural mechanics and theoretical physics. For that reason a reference to selected sources has nothing to offer for its better comprehension and is omitted.

I apologize for any extent that the Greek syntax is intruding through the English text. This was a conscious choice aiming to avoid spending too much time for a proper complete linguistic adaptation, after taking into consideration that the clarity is hardly lost in a text, which is dominated by a scientific terminology.

Acknowledgments for their assistance is owed to our secretary Mrs. Vetta Halepli, who was typing and retyping manuscripts during some decades, with many thanks for her diligence and devotion· also to our CAD operators Mrss. Dora Vasilaki, Ioanna Papadopoulou, Maria Panagopoulou and Mr. Petros Drosos for the diligent reproduction of the figures· to my son Emmanouil, colleague and collaborator, who was charged to impose order to the chaos of the continuously revised matter up to its final formation· to my daughter Orianna, for her help with the final touches of the book and my wife Maria for her constant support and patience.

June 2021,

D.E. Bairaktaris

## INTRODUCTION

The quantum characteristics of the elementary particles have been derived from the observation of their behaviour during their interactions with other particles or with the measuring devices, which take place in a space free or influenced by imposed magnetic or electric fields. In this view the particles are considered as rigid material or massless points of infinitesimal size. What is referred as the radius of a miniature spherical shape is actually the distance at which the scattering with other particles occurs. The behaviour is considered as rigid motion, which is described by two entities, the one consisting of a set of kinematic parameters containing the translations, the rotations, the orbital and the angular velocities, which define their trajectories, and the other consisting of the set of the conjugated dynamic parameters, which contains the forces, the moments and the orbital and the angular momentums. Each one of the conjugated kinematic and dynamic entities is represented by a pair of a polar and an axial vector and is defined by the set of their six vector components in a 3-dimensional reference system. The sum of the dot products of vectors representing the dynamic entities with those of the conjugated kinematic ones represents both, the work, which is equal to the variation of the potential energy, and the kinetic energy of the particle, according to the following arrangement:

| **Conjugated Entities** | **Work** | | **Kinetic Energy** | |
|---|---|---|---|---|
| | **Polar** | **Axial** | **Polar** | **Axial** |
| Dynamic | Force | Moment | Orbital momentum | Angular momentum |
| Kinematic | Translation | Rotation | Orbital velocity | Angular velocity |

The system of the 2x3 polar and 2x3 axial vector components of the kinematic entities, in a 3D reference system constitutes the configuration space of the particle, while the system containing moreover the polar and the axial vector components of the conjugated dynamic entities constitutes its phase space. The attribution of polar vectors, representing intrinsic moments or angular momentums in particles of infinitesimal size, which imply properties of polarization at distinct directions, constitutes the main difference from the considerations concerning the systems of material points in classical mechanics and the source of the quantum characteristics.

Although the assumption of the infinitesimal size permits the classification of the elementary particles with respect to their quantum characteristic, does not offer the capacity for the natural origin of these characteristics to be interpreted. The origin has to be searched to the internal structure of the particles, something implying a finite size, instead of the infinitesimal one, to be considered, under the additional assumption for the substance contained to possess the properties of a deformable continuum. The investigation of the deformation's modes and their relation with the external behavior of the particle, constitutes the main subject of this treatise. To that purpose the principles and methods of the continuum mechanics have to be taken into account. Their adaptation to the methodology, which is established for the specific case of the particles, it's the subject of chapter 2.
Some basic ideas of continuum mechanics have to be mentioned. The volume of a body is divided in elementary cubic segments. Each one elementary segment undergoes inertial mass polar vector forces or momentums throughout its volume and normal and shear strains at the faces due to its interaction with the adjacent segments. The segments having faces on the boundary surface of the body undergo and the polar and axial vector forces or momentums due to the interaction with other bodies.

An explicit definition for the concept of the particle is necessary. By this term a set of material points of any size, macroscopic or microscopic, is meant, where the behavior of each one point is defined by a finite set of parameters which characterize the state of the whole system. Examples of particles according to this definition are the rigid bodies, the transversal slices of elastic beams with infinitesimal thickness, the vortices developed in fluids, as well as the elementary particles of the quantum physics.

The set of the possible modes of deformation, which define completely the total state of the particle and constitute its dimensions in the configuration and in the phase space, result from the application of the *Octal Decomposition* procedure, which is described in detail in chapter 3. Two sets, each one containing eight modes are derived from this procedure, with discriminating property the conservation or not of the right angles after the deformation, which are referred as *conformal* and *contra-formal* respectively. The sum of the work or the kinetic energy of the polar vector forces and momentums, which are applied to the totality of the elementary segments of the particle, for the conjugated with them translations and velocities corresponding to a mode, results equal to the product of a single dynamic by a single kinematic entity, which constitute the state parameters of the particle corresponding to this mode. The state parameters have different representations throughout the eight modes, conformal and contra-formal. They consists of the three components of a polar vector in a 3D orthogonal reference system, the three components of an axial vector in the same system and two scalars, a real one and a triple vector product.

The dynamic and kinematic parameters for the total state of the particle are presented by the sum of the eight components in an eight dimensional space with a basis consisting of a real scalar unit, three real polar unit vectors in a 3-dimensional orthogonal reference system, and an imaginary scalar unit equal to the triple vector product of the three polar vector units. This basis is isomorphic to the octonion one and the product between two eight-folds, consisting of the sum if the conformal or the contra-formal dynamic and kinematic state parameters of the particles, satisfies the non-commutative and the non-associative – alternative laws governing the algebra of the octonions. The consequences of this isomorphism with the octonions' algebra and analysis, which implies the distinct branches of the real, the complex and the vector algebra and analysis to be unified a single mathematical system, is the subject of chapter 1 and constitutes the mathematical tool for the description and the investigation of the internal structure of the particles. Moreover unified mathematics will be possible proved to be a prerequisite to the purpose of deriving unified theories in physics.

The total set of the sixteen conformal and contra-formal state parameters is furthermore considered as the sum of two similar subsets. The one contains the dynamic state parameters which are representative of the forces or momentums applied on faces of the elementary segments lying on the external surface of the particle, together with the conjugated with them kinematic ones. They are due to the interaction with other particles of with external fields and are referred as *interacting*. The main characteristic is their observability from detecting and measuring devices. The second subset contains the dynamic state parameters which are representative of the forces or momentums applied on the faces of the elementary segments lying in the interior of the particle and on their volume and the conjugated with them kinematic ones. They are referred as *hidden*, as they are not observable.

Nevertheless their values are derived from the conditions which relate them with those of the interacting state parameters.

The necessity for this distinction between interacting and hidden state parameters becomes apparent in the case of interactions between particles belonging to 1-, 2- and 3-dimensional aggregates, which constitute the subject of the chapters 4, 5 and 6.

Finally, a 32-demensional configuration space results, which is referred to four 8-dimensional similar octonion bases, corresponding to the conformal and the contra-formal representations of the interacting and the hidden state parameters. The sic interacting parameters, which are represented by conformal polar and axial vector components, correspond to rigid body motions of the particle and define the variation of its position and orientation in the external space. For that reason the corresponding to them six dimensions are considered as external. Moreover they are representative of systems of forces or momentums with non-zero vector sum. The rest 16-6=10 interacting, or 32-6=26 total state parameters correspond exclusively to modes of internal deformation of the particle, which are associated with systems of forces of zero vector sum, although they can be represented by non-zero polar or axial vector state parameters. The corresponding to them dimensions in the configuration space, where the characteristics of the internal structure of the particle are referred, are considered as internal dimensions. The numbers 10 or 26 of the internal dimensions coincides with those derived from the Super Strings and the M-theories for the dimensions of the strings.

The real scalar state parameters represent a converging or diverging throughout the volume linear deformation of the elementary segments, expansion or contraction, which is conjugated with a likewise converging or diverging positive or negative pressure applied on them. In a similar manner the imaginary scalar state parameters correspond to sets of forces or momentums applied on the elementary segments with an appropriate distribution throughout the volume of the particle, which implies the development of spherical distortional moments or angular momentums, already referred as *tri-moments* in fluid dynamics and as *bi-moments* in structural mechanics of elastic beams. The imaginary scalar dynamic parameter is proved to represent the real form of the spin in conditions free of external influences. According to the present consideration, in the presence of magnetic fields, where the spin is experimentally detected and measured, the spherical distortional angular momentum is actually transformed to the axial vector form of the angular momentum.

A reference system for the cartography of the Particles' World is completed. And a challenge for explorations is offered.

## OVERVIEW

The main subject of the present treatise is contained in the chapter 3, where the concept and the state parameters defining the particles in their elementary form are presented. It is suggested for a first reading, together with the part 1.1 of chapter 1, concerning the algebra of octonions in the form appropriate for the presentation of physical entities. The chapter 2 is introductory for the rest chapters 4, 5 and 6 which concern mainly extended consideration in subjects of structural mechanics, beams and surface structures, resulting to the discovery of the excess state parameters beyond the six, which are taken into consideration in the present day theory. Nevertheless the state parameters, which are defined in chapter 4 for the beams, appear an interesting similarity with the internal dimensions, which are predicted from the theories of the superstrings.

An abridged presentation of the content of the different chapters is contained on the following sections.

### 1. Octonion Mathematics for Physics

In the subsequent chapters of the present treatise will become clear that the octonion mathematics consists the appropriate mathematical tool offered for the description of the complete set of entities and relations permitting an advanced investigation of the physical properties attributed to deformable continuous media. The octonion algebra was initially introduced by Graves (1843) and Cayley (1845) as an extension in $2^3$ dimensions of the $2^2$ – dimensional quaternion algebra of Hamilton. Since then no important applications of this algebra are referred in theoretical treatments concerning topics of physics. Only recently group theoretical properties of the octonion algebra have been associated with the basic symmetries governing the gauge fields and supersymmetry theories for the elementary particles.

In the present formulation the octonion is considered as a set of eight linearly independent components containing all kinds of mathematical objects already used for the representation of the entities of physics, constituting the subject of separate mathematical topics, namely those of real and imaginary scalar fields and of 3-dimensinal polar and axial vector spaces. Their inclusion in the octonionic structure implies the possibility for the definition of inverse elements, which are not provided for ordinary vector spaces, and thus permitting their treatment as elements of a division algebra. The operations applied on the octonions are extensions and unifications of the operations defined for the parent mathematical topics accompanied by the complexity due to the noncommutativity and non-associativity - alternativity of the octonion algebra.

The class of the Linear Algebras without Divisors of Zero (LAwDoZ), or Linear Division Algebras, contains four members· the Real (R), the Complex (C), the Quaternion (H) and the Octonion ($\omega$) algebras. Each element of a LAwDoZ consists of a number of linearly independent components consisting of real and imaginary entities. The commutative, the associative and the distributive laws are enforced for the addition, but not for the multiplication between elements in the totality of the members of the class.

According to a theorem of Frobenius (1878) concerning the first three members, which was completed later by Hurwitz (1898) for the totality of the members of the class, there is not the possibility for the creation of other linear division algebras obeying to the prescribed for each one combination of compliance or not with the commutative and the associative laws. The same result was verified later by Radon, Jordan, von Neumann, Wigner, Freaudenthal, Chevaley, Dickson, Albert and Cartan, on the basis of the group representation. Artin (1925) noticed the alternativity property for the octonions and Milnor (1958) showed that the creation of linear division algebras with elements consisting of more than eight components is not possible.

The content of this chapter is restricted mainly to the presentation of the selected and proposed notation for the octonions and their components, the corresponding algebraic expressions as well as the detailed execution rules for the basic and the composite operations with octonions for application in the algebra, the geometry and the analysis. With the emphasis given to the creation of an active mathematical tool aiming at the derivation of formulas of physics in octonionic expressions, extensive proofs of the basic theorems, already cited in the existing literature on the subject are omitted.

The main innovations introduced by this revised formulation of the octonion mathematics are the following:

- The algebra of quaternions contains three, distinct but equivalent, imaginary scalar units, which in the present literature on the subject are denoted by the symbols $i, j, k$, with $i^2 = j^2 = k^2 = -1$ and $ij = k,\ jk = i,\ ki = j$. From another point of view these imaginary units, considered as linearly independent unit imaginary vectors in the directions of the axes of an orthogonal reference system, are elements of the *quaternion group* which is defined by the operation

$$c = a \cdot b + (a \times b)$$

where $(a, b, c)$ is any permutation of $i, j, k$. But the inclusion of these elements to the same group does not necessarily imply that they are of the same type. From the Cayley - Dickson procedure of the field of quaternions from a complex field $x + iy$, where $x, y$ instead of reals, are replaced by elements of another complex field

$$x = a + jc, \quad y = b + jd, \quad (j \neq i,\ \ j \cdot i = 0)$$

the quaternions take the form

$$q = x + iy = a + jc + i(b + jd) = a + ib + jc + (ij)d$$

From the above relations results

$$ij = k = i \times j$$

so that, for $i, j$ *polar* vectors, $k$ is an *axial* vector perpendicular to both $i$ and $j$. In order to be avoided any confusion with $i, j, k$, introduced usually as numerical indices in the equations, the unit polar vectors are denoted by low right indices and the axial vectors by upper right indices, taking the notation

$$i = e_x,\ \ j = e_y,\ \ k = e^z$$

where $x, y, z$ are the directions of the orthogonal reference system. By this formulation is raised the peculiarity of the inclusion of three identical scalar imaginary units in the same field, which differ solely in notation.

- From the similar generation procedure of the field of octonions from quaternions having the revised form

$$a + e_x b + e_y c + e^z d$$

after replacing the real coefficients $a, b, c$ and $d$ by elements of a third complex field with unit imaginary polar vector $e_z \neq e_x, e_y$, so that for real $a_i (i = 1 \; to \; 8)$,

$$a = a_1 + e_z a_s, \quad \text{b}{=}a_2 + e_z a_6, \quad c = a_3 + e_z a_7, \quad d = a_4 + e_z a_8,$$

the octonion takes the form

$$\omega = a_1 1 + e_x a_2 + e_y a_3 + e^z a_4 + e_z a_6 + (e_x e_z) a_6 + (e_y e_z) a_7 + (e^z e_z) a_8,$$

According to the relation above the two first products contained in parentheses are axial vectors of the form

$$e_x e_z = e_x \times e_z = -e^y, \qquad e_y e_z = e_z \times e_z = e^x$$

while the third one is the triple vector product

$$e^z e_z = (e_x e_y) e_z = (e_x \times e_y) \cdot e_z = e$$

equal to the scalar imaginary unit, which furthermore will be denoted by the symbol $e$, instead of the traditional $i$. The same symbol is used as upper right index denoting imaginary scalar entities, while the real scalar entities are denoted by the lower right index 1. After replacing the above relations, the octonions take the form

$$\omega = a_1 1 + a_2 e_x + a_3 e_y + a_4 e^z + a_5 e_z - a_6 e^y + a_7 e^x + a_8 e$$

Under this configuration the set of the eight components of the octonions consists of one scalar, three polar vector components, three axial vector components and an imaginary scalar, so that the octonion can be represented by the concise form

$$\omega = \omega_1 + \omega_p + \omega^a + \omega^e$$

where $\omega_p$ and $\omega^a$ represent a polar and an axial vector respectively.

- The product of two octonions is expressed as a sum of partial products of scalar by scalar, scalar by vector, dot and cross products of polar or of axial vectors and moreover the corresponding to them newly introduced *hole* and *grid* products of polar by axial vectors.

- Because of the non-commutativity and the non-associativity – alternativity properties for the multiplication between octonions the result of a product between more than two octonions depends on the sequence of execution of the operations. The sequence is denoted by parentheses containing two-term products consisting of single octonion components or two-term products of lower order. Outside the parenthesis of each one two-term products is assigned a lower or upper right index corresponding to the indices of the pair of the included terms according to the multiplication table for octonions, considered thus furthermore the two-term products as a single octonion component.

- In algebraic expressions of general form, where the eight indices of the octonion components are denoted by the general indices $I, J, K, \ldots.$, the final sign due to the effect of the non commutativity is denoted by the introduction of the two index symbol $\vartheta_{IJ} = 2\delta_{IJ} - 1$, where $\delta_{IJ}$ is an extension of the Kronecker's symbol for octonions. The final sign of an expression due to the effect of the non-associativity – alternativity property is denoted by the introduction of the three index symbol $\vartheta_{IJK}$.

- Except the concepts of the conjugate and the inverse of an octonion is introduced the concept of the *reverse octonion* having components the inverses of the components of the parent octonion. The polar and the axial vectors contained in the reverse are not generally collinear to those contained in the parent octonion, except the cases they are collinear with one of the coordinate axes or their three coordinates have equal values and they form equal angles with the three coordinate axes. As collinearity between parent and reverse vectors is significant for the definition of some basic scalar entities in continuum mechanics, with values independent of the direction of the reference system, a specific type of orthogonal transformation is introduced which transfers any vector to the equiangular position with respect to the axes of the reference system.

- A free polar vector is applied at a point defined by its position vector. By a first orthogonal transformation the position vector is transferred to the equiangular position. By a second orthogonal transformation, consisting to a rotation around the position vector, the $z$-axis is transferred on the plane of the two vectors, the free polar and the position one. The normal to the position vector at its end point on that plane intersects the new $z$-axis at a point which is referred as the *pole* of the given pair of vectors. The pole, the position vector and the normal to it at its end point transverse vector up to the pole constitute the *polar system* of any given pair of vectors. The polar system is proved indispensable for the definition of real and imaginary scalar resultants of a polar vector field.

- A closed system of points interconnected by internal geometrical constrains is defined through the reference of the points contained in it in a local coordinate system $x', y', z'$. The state of the system is determined by a set of external and a set of internal state parameters represented by octonion components. The set of the external state parameters consists of the polar position vector of the origin of the local reference system with respect to the global $x, y, z$ and the axial orientation vector having direction and length the axis and the angle of rotation respectively which correspond to the orthogonal transformation from the local to the global system. The real and the imaginary scalar components belong to the set of the internal state parameters and represent a uniform dilatation and a uniform twist respectively of the internal constrain between the points of the system.

- In the domain of the Octonion Analysis the definition of the octonion function is initially introduced considered as an octonion with components taking values depending on the values of the components of octonion arguments or variables.

- The partial derivative of an octonion faction with respect to an argument component is defined, equal to the sum of the products of the infinitesimal variation of the components of the function due to an infinitesimal variation of the argument component by the reverse of the argument variation. Moreover the derivation rules are presented concerning products of octonion functions and higher order derivatives of an octonion function.

- The concept of the analytic functions of complex variables is extended to the analytic octonion functions with argument a complete octonion. The necessary and sufficient condition for an octonion function to be analytic is expressed by a system of 56 partial differential equations between the partial derivatives of the real coefficients of the octonion function components and constitutes an octonion extension of the Cauchy-Riemann system of equations concerning complex analytic functions.

- From the four terms expression of the octonion function and the octonion argument results the corresponding four terms expression of the derivative. Each term of the derivative expression can be presented as an octonion dot or cross or hole or grid product of the function by a single differential operator, which is the octonion extension of nabla differential operator defined for real functions. Under this form an analytic octonion function can be considered as an octonionic potential of an octonion field, with the four terms of the derivative constituting the real divergence, the gradient, the curl and the imaginary divergence respectively. Similarly the complete expression of the derivative can be presented as the product of the function by a total differential operator composed of the sum of the partial operators corresponding to each one term of the derivative.

- The integral of an octonion function is defined along an octonion path with components expressed by functions of a single scalar parameter as variable.

- The Canchy's integral theorem for complex analytic functions is extended to the octonion analytic functions. From this theorem, after the introduction of the concept of the *primitive reverse octonion*, the extended form results of the Canchy's integral formula for octonion analytic functions.

## 2. The Deformable Continuum

The present treatise is governed by the basic assumption that the elementary particles in their microstructure, instead of being considered as miniatures of spherical rigid bodies accompanied by a wave function, they have shape, size and consist of a deformable material. For that reason, under the action of forces and momenta, except of their rigid body motion in the space, deterministic or probabilistic, they execute additionally deformation of their shape due to internally developed interacting forces between their constituent material points. Concepts and relations already established in the mechanics of the continuous media, solids and fluids, expressed in the octonion notation are repeated in order to be used in the subsequent chapters 4 and 5 concerning the treatment of the aggregates of particles.

The forces per unit area, which in overall static condition are developed at the interfaces between the material points of infinitesimal size and are referred as *stresses*, are conjugated with the corresponding to them relative longitudinal and transversal displacements per unit length of the their spacing, which are referred as *strains*. The product of the stresses by the strains is the work done per unit volume due to the deformation, which is equal to the dynamic energy per unit volume stored in the material. In a similar manner the conjugate concepts are introduced of the mementa per unit area transferred between adjacent material points during their relative motion, which are referred by the term *impulses* and the corresponding to them longitudinal and transversal relative velocities, which are referred by the term *strain rates*. The product of the impulses by the strain rates equals the internal kinetic energy acquired by the material.

## 3. The Particle

As particle is considered a set of material points interconnected through massless bonds with specified mode of interaction. Because of their infinitesimal size only polar and not axial vector forces or momenta are applied on them, having a total resultant force or momentum with respect to a reference center, furthermore referred as total composant, which is equal to the sum of the isolated polar vector forces and momenta, each one multiplied by an appropriate *lever-arm* function of the coordinates of the corresponding material point with respect to an orthogonal reference system, having its origin at the center.

The defining property of the particle consists to the condition the displacement or the velocity of each one material point due to the action of the set of forces and momenta applied on them, result from a single displacement or velocity, parameter, referred furthermore as *motive*, which is unique for the whole particle, multiplied by a *shape* function of their coordinate, equal to the corresponding lever-arm function.

The introduced Octal Decomposition procedure consists to the replacement of each one polar vector force or momentum applied on the material points by the superposition of eight equivalent sets of eight forces or momenta applied at points lying in the eight quarters of the reference system in positions symmetrical to the parent point with respect to the reference planes. A component of the total composant corresponds to each one of the eight sets. The set of the resulting eight components, furthermore referred as *composants*, contains three polar and three axial vectors, both in the directions of the axes of the reference system, a real and an imaginary scalar and constitute the components of an octonion, which represents the total composant. The polar system associated with the polar vector force applied on the material points has a decisive contribution to the formation of the corresponding composants.

The components of a similar octonion represent the motives corresponding to the composants. The motives and the corresponding composants are furthermore divided into two constituent parts, a conformal and a contra-formal, the first one characterized by the conservation of the orthogonality. Finally the state of the particle is defined by two pairs of octonions, a conformal and a contra-formal, corresponding to the pairs of the conjugate entities of the composants and the motives. The sic polar and axial vector composants and motives of the conformal octonions correspond to rigid body motion of the particle and constitute the external dimensions of its phase space. The set of the rest ten components, containing the two scalar ones of the conformal octonions, real and imaginary, as well as the eight components of the contra-formal octonions, correspond to deformation modes of the particle and constitute the internal dimension of its phase space.

Each one polar and axial vector composant represents force or momentum and moment or angular momentum respectively, while its conjugate polar and axial vector motive represents translations or orbital velocities and rotations or angular velocities respectively. The real scalar composant represents the internal pressure, isotropic or anisotropic, while the conjugate real scalar motive represents the associated expansion or contraction. The entity known as Tri-moment in Fluid Mechanics or Bi-moment in Structural Mechanics is represented by the imaginary scalar composant. The hypothesis that the spin of the elementary particles in conditions free of external effects is represented by the imaginary scalar composant, is proved to be compatible with its form of the angular momentum, which is detected in the presence of magnetic fields.

### 4. One-dimensional Aggregates of Two-dimensional Particles

The beams, which constitute a prominent subject of Structural Mechanics, is a representative model for this type of aggregates. Transversal slices of infinitesimal thickness along the axis of the beam are considered as the constituent particles of the aggregate. Adjacent particles interact through one normal and two transversal shear stresses, which are developed between the material points lying of the two opposite cross sections of the beam at the interfaces between them.

Instead of the octal decomposition procedure, which was adopted for the description of the free particles, two separate Quaternal decompositions are introduced for the interconnected particles of the beam. The one of them is applied on the forces due to the normal stresses acting in the longitudinal direction and results to composants represented by two quaternions, one conformal and the other contra-formal, both multiplied by the imaginary unit. The second one is applied on the forces due to the shear stresses acting in the transversal direction and results to composants represented by two additional conformal and contra-formal quaternions. Furthermore is proved that each one shear quaternion is identified with the derivative of the relevant normal quaternion with respect to the coordinate along the axis of the beam, forming by this property a Generation Rule between composants. The two pairs of the conformal and the contra-formal quaternions are composed in two octonions, a conformal and a contra-formal respectively, which represent the totality of the interacting composants between adjacent particles.

The six polar and axial vector composants contained in the conformal octonion, are associated with motives representing rigid body motions of the particle, three translations and three rotations and correspond to its external dimensions. The remaining ten composants, consisting of the two conformal real and imaginary scalars and the totality of the eight contra-formal, are associated with motives representing modes of deformation and correspond to its internal dimensions.

Except the three kinds of stresses, which are developed at the interfaces between the particles, three more kinds are developed in their interior the two faces, two normal in the transversal directions and one shear in the longitudinal direction, implying the development of the relevant strains and the corresponding to them dynamic energy stored in the material. A set of sixteen additional composants correspond to the forces due to these stresses, which are represented by two octonions, a conformal and a contra-formal one, referred furthermore as *Hidden Composants*. They are associated with motives representing exclusively modes of deformation and belong therefore to the internal dimensions of the phase space of the particle. The hidden composants are not observable and detectable, because the observation presupposes the interaction with external observers. They are defined from the observable interacting composants, after applying successively the generation rule which is established between normal and shear quaternions.

Finally results that the state of particles interconnected in a one-dimensional aggregate is defined by 32 pairs of conjugate composants and motives, which are contained in two pairs of conformal and contra-formal octonions, one for the interacting and the other for the hidden entities. Only 6 of them belong to the external dimensions of the phase space, the 3 polar and the 3 axial vector conformal interacting composants. To the rest 26 internal dimensions belong the 2 conformal, the 8 contra-formal and totally 10 interacting, as well as the 16 hidden composants and motives. The coincidence of the 10 and 26 internal dimensions with the relevant numbers predicted from the superstrings and M-theory, is compatible with the consideration of the superstrings as one-dimensional aggregates.

Generally in Physics and especially in Structural Mechanics only the entities belonging to the six external dimensions are taken into account. For the last case the omission of the internal ones is justified from the fact that the corresponding to them dynamic or kinetic energy constitutes a negligible part of the total energy of the system. This is not certified for the case of Particle Physics.

## 5. Two-dimensional Aggregates of One-dimensional Particles

The membranes, the plates, the shells and generally the surface structures, which belong to the subject of Structural Mechanics, constitute representative models of this type of aggregates. They are defined by a reference surface and the thickness in the direction transversal to it, which is a function of its points. As particles are considered one-dimensional threads along the thickness corresponding to the points of the reference surface and consist of material points which are interconnected with those of the adjacent particles. On the material points are applied polar vector forces consisting of the superposition of those due to the stresses developed between adjacent particles and the external forces applied on them.

The composants of the particles result from independent dual decomposition procedures of the forces applied on the material points, each one concerning their component in one of the three directions of their local reference system. Because of the infinitesimal dimension of the transversal cross section of the particles, the axial vector composants with respect to their longitudinal axis, as well as the imaginary scalar one are omitted as higher order infinitesimals. The dual decomposition procedure is executed in the one-dimensional space of the longitudinal axis of the particles, where the polar systems are identified with the reference one, therefore making impossible the formation of contra-formal composants.

Furthermore each one composant is divided in three distinct parts, the one containing the external forces and the other two the forces due to the normal and shear stresses respectively. By this division the internal forces result, which are provided in the theory of the surface structures, except the real scalar di-force in the direction of the thickness, this is ignored in the present day status of the theory, because of its negligible effect in the total energy of the system. The axial internal force in the direction of the thickness as well as the di-force corresponding to the external forces, both due to normal stresses in the same direction, which don't interact with stresses of adjacent particles, are hidden forces, while all the rest internal forces are interacting.

## 6. Three-dimensional Aggregates of Particles

The deformable continuum, which is presented in chapter 2, is an aggregate of particles of infinitesimal size, where as particles are considered the material points subjected external and interacting polar vector forces.

As aggregates constricting of solid particles with finite size interconnected through bonds of different DOF, are considered the molecular, the crystal, the granular, as well as the composite massive natural and technological structures. The particles are subjected external moments or momenta as well as the interacting stresses or impulses developed at their contact interfaces, which are represented by pairs consisting of a conformal and a contra-formal octonion. The state of each one particle is defined by 16 composants, which constitutes the components of a conformal and a contra-formal octonion and represent likewise the hidden forces which correspond to the stresses developed in its interior.

As particles of infinite size are considered mass or electric charge points accompanied by their gravitational or electrostatic fields, sources, sinks or vortices with the relevant to them fields of flow, which are developed in fluids, liquids or gases etc. The determination of the state for each one particle is the subject of the *many body problem*, which in Physics appears known theoretical difficulties. Their treatment is restricted here to the presentation of the condition to be satisfied for the points of surfaces separating the influence regions of the single particle, where the displacements or the velocities acquire minimum or maximum values.

# 1. OCTONIONS

## 1.1 OCTONION ALGEBRA

### 1.1.1 Linear Algebras without Divisors of Zero (LAwDoZ)

The structure of each one algebra has as follows:

**Real Algebra (R)**

Contains elements $\alpha_R$ consisting of one real scalar component, unit element 1 and the commutative associative and distributive laws enforced for multiplication. Moreover the elements of R constitute an ordered set.

**Complex Algebra (C)**

Contains elements $\alpha_c$ consisting of two linearly independent components $\alpha_1$ and $\alpha_x$ referred to a basis of two linearly independent unit elements $e_1$=1 and $e_x$ according to the relation

$$a_c = \hat{a}_1 + a_x = \hat{a}_1 1 + \hat{a}_x e_x \tag{1.1.1}$$

where

| | | |
|---|---|---|
| $\hat{a}_1, \hat{a}_x$ | : | real scalar coefficients |
| 1 | : | the real scalar unit ($1^2 = 1$) |
| $e_x$ | : | an imaginary unit polar vector ($e_x^2 = e_x e_x = -1$) |

The above notation for the real scalar coefficients will be justified in Section 1.1.5.
In complex algebra the three laws are likewise enforced for multiplication but the set of elements $\alpha_c$ is not ordered.

**Quaternion Algebra (H)**

Contains elements $a_H$ consisting of four linearly independent components $a_1, a_x, a_y, a^z$, referred to a basis of four linearly independent unit elements $1, e_x, e_y$ and $e^z$, according to the relation

$$a_H = a_1 + a_x + a_y + a^z = \hat{a}_1 1 + \hat{a}_x e_x + \hat{a}_y e_y + \hat{a}^z e^z \qquad (1.1.2)$$

where beyond the already defined in (1.1.1) symbols, the rest represent the following entities :

| | | |
|---|---|---|
| $\hat{a}_y, \hat{a}^z$ | : | real scalar coefficients |
| $e_y$ | : | an imaginary unit polar vector orthogonal to $e_x\left(e_y^2 = e_y \cdot e_y = -1, e_x \cdot e_y = 0\right)$ |
| $e^z$ | : | $a$ real unit axial vector or pseudo-vector, orthogonal to both $e_x$ and $e_y$ in a right-handed direction $\left(e^z = e_x \times e_y,\ e^{z2} = -1\right)$ |

Generally in the applied hereafter notation the right upper indices define real scalar and axial vector components, while the right low indices define imaginary scalar and polar vector components.

A quaternion $a_H$ can be considered as a complex element in a new basis of unit elements 1 and $e_y$ with the scalar coefficients replaced by complex elements in the basis of unit elements 1 end $e_x$ according to the relation

$$a_H = (a_1 1 + \hat{a}_x e_x)1 + \left(\hat{a}_y 1 + \hat{a}^z e_x\right)e_y = \hat{a}_1 1 + \hat{a}_x e_x + \hat{a}_y e_y + \hat{a}^z\left(e_x e_y\right) \qquad (1.1.2a)$$

which coincides with (1.1.2) for $e_x e_y = e_x \times e_y = e^z$

In quaternion algebra the associative and distributive laws are enforced for multiplication, but not the commutative law, which follows the relation

$$e_I e_J = \theta_{IJ} e_J e_I \qquad (1.1.3)$$

where,

$I, J$: any two of the above four right indices of the quaternion components considered as different for the upper and the low position,

$$\theta_{IJ} = 2\delta_{IJ} - 1 = \begin{cases} 1 \text{ for } e_I = \pm 1, \text{ or } e_J = \pm 1, \text{ or } e_I = \pm e_J \\ -1 \text{ for } e_I \neq \pm 1, \text{ and } e_J \neq \pm 1, \text{ and } e_I \neq \pm e_J \end{cases}$$

$$\delta_{IJ} = \begin{cases} 1 \text{ for } e_I = \pm 1, \text{ or } e_J = \pm 1, \text{ or } e_I = \pm e_J \\ 0 \text{ for } e_I \neq \pm 1, \text{ and } e_J \neq \pm 1, \text{ and } e_I \neq \pm e_J \end{cases}$$

the extended Kronecker symbol.
The axial unit element $e^z$, fulfills the identity $e^{z2} = -1$, resulting from the following procedure:

$$e^{z2} = (e_x e_y)(e_x e_y) = -(e_x e_y)(e_y e_x) = -\left((e_x e_y)e_y\right) e_x = -(e_x e_y^2)e_x = e_x^2 = -1$$

In the upper left quarter of the Table 1.1.1a is contained the multiplication table between quaternion unit elements, where the element corresponding to the product $e_I e_J$ lies at the intersection of the $I$ −row with the $J$ −column.

**Octonion Algebra ($\omega$)**

Contains elements $a_\omega$ consisting of eight linearly independent components $a_1, a_x, a_y, a^z, a_z, a^y, a^x, a^e$ referred to a basis of eight linearly independent unit elements1, $a_x, e_y, e^z, e_z, e^y, e^x,$ and $e$, according to the relation

$$a_\omega = a_1 + a_x + a_y + a^z + a_z + a^y + a^x + a^e = \hat{a}_1 1 + \hat{a}_x e_x + \hat{a}_y e_y + \hat{a}^z e^z + \hat{a}_z e_z + \hat{a}^y e^y + \hat{a}^z e^x + \hat{a}_e e \quad (1.1.4)$$

where, beyond the already defined in (1.1.1) and (1.1.2) symbols, the rest represent the following entities:

$\hat{a}_z, \hat{a}^y, \hat{a}^x, \hat{a}_e$ : real scalar coefficients

$e_z$ : an imaginary unit polar vector forming with $e_x$ and $e_y$ a right-handedorthogonal system $\left(e_z^2 = e_z \cdot e_z = -1, e_x \cdot e_z = e_y \cdot e_z = 0\right)$

$e^y, e^x$ : unit axial vectors or pseudo-vectors forming with a left-handed orthogonal $\left(e^y = e_x \times e_z,\ \ e^x = e_y \times e_z,\ \ e^{y^2} = e^{z^2} = -1\right) e^z$

$e$ : the imaginary scalar or pseudo-scalar unit equal to the triple scalar product of the polar unit vectors $\left(e = e^z \cdot e_z = \left(e_x \times e_y\right) \cdot e_z, \quad e^2 = -1\right)$

By repeating the generation procedure already applied for the quaternions in the relation (1.1.2a), an octonion $a_\omega$ can be likewise considered as a complex element in a new basis of unit elements 1 and $e_z$ with the scalar coefficients replaced by quaternions defined in the basis of unit elements $1, e_x, e_y, e^z$ according to the relation

$$a_\omega = (\hat{a}_1 e_1 + \hat{a}_x e_x + \hat{a}_y e_y + \hat{a}^z e^z) \cdot 1 + (\hat{a}_z e_1 + \hat{a}^y e_x + \hat{a}^x e_y + \hat{a}_e e^z) e_z$$

which coincides with (1.1.4) for

$$e_x e_z = e_x \times e_z = e^y, \qquad e_y e_z = e_y \times e_z = e^x, \qquad e^z e_z = e^z \cdot e_z = e$$

From the nature of their contribution to the generation procedure the octonion components $a_1, a_x, a_y, a_z$ and the corresponding to them unit elements $1, e_x, e_y, e_z$ are characterized as *primitive*, while the rest $a^x, a^y, a^z, a^e$ and the corresponding unit elements $e^x, e^y, e^z e$ are characterized as *derived.*

In the multiplication table 1.1.1a the unit element corresponding to the product $e_I e_J$ lies at the intersection of the $I-row$with the $J-colunm$.

where, $I, J, K$: any three of the above eight right indices of the octonion components considered as different for the upper and the lower position

From the relation (1.1.4) results that any octonion $a_\omega$ can be represented by a sum

$$a_\omega = a_1 + a_p + a^a + a^e \qquad (1.1.5)$$

where

| | | |
|---|---|---|
| $a_1$ | : | a real scalar |
| $a_p = a_x + a_y + a_z$ | : | a 3-dimersional imaginary polar vector |
| $a^a = a^x + a^y + a^z$ | : | a 3-dimesional axial vector or pseudo-vector |
| $a^e$ | : | an imaginary scalar or pseudo-scalar |

For the arrangement of the unit elements according to the relation (1.1.6) is applied the multiplication table (1.1.1.b)

The polar vectors $a_p$ are referred to a right handed reference system with basis the unit vectors $e_x, e_y, e_z$ while the axial vectors $a^a$, as defined in (1.1.4), results to be referred to a left-handed reference system with basis the unit vectors $e^x, e^y, e^z$ (Fig. 1.1.1a). The existence of two counterpoising reference systems does not permit to create formulas containing general indices $i, j, k$ representing in that order any cyclic permutation of the indices $x, y, z$ considered likewise in that order. The main deficiency results from the fact that in the left-handed system it is $e^y = e_x e_z$, while a definition $e^y = e_z e_x$, corresponding to the objective for obtaining formulas containing cyclic permutations of the indices, is compatible with the reference of $e^x, e^y, e^z$ in a right-handed system. For that purpose in the sequence both polar and axial vectors will be referred in two identpointing right-handed reference systems (Fig. 1.1.1b), with the appropriately modified multiplication tables 1.1.1c and 1.1.1d being applied in that case. From these the multiplication tables 1.1.1e and 1.1.1f result which contain general indices $i, j, k$, representing in that order any cyclic permutation of the initial indices $x, y, z$, considered likewise in that order.

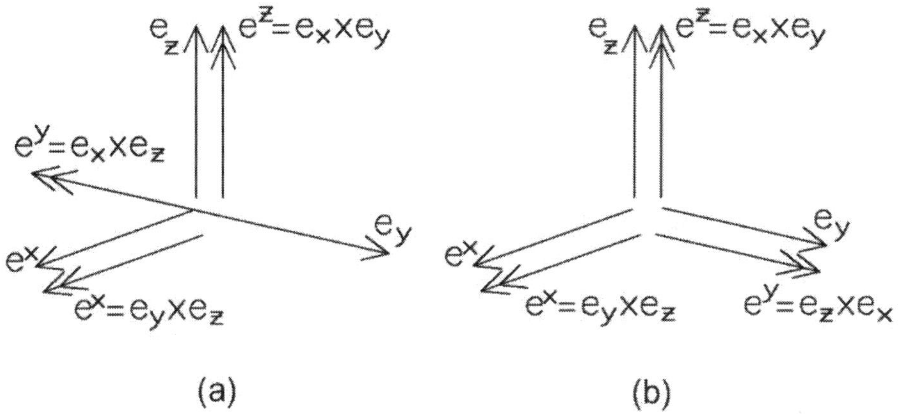


*Figure 1.1.1 - Counterpointing (a) and identpointing (b) reference systems*

From the above presentation results that the imaginary unit element $\ell$ constitutes an element of the octonion basis and not of the complex basis, where the second element consists of an imaginary unit polar vector. Moreover the multiplication by $\ell$ transforms polar to axial vectors and vice-versa, as well as identpointing pairs of reference systems to counterpoising ones and vice-versa, and constitutes for that reason a parity transformation.

The octonion algebra is non-commutative and non-associative – alternative. For the non-commutativity is applied the relation (1.1.3) for indices $I,J$ running over the eight indices of the octonion components. For the non-associativity – alternativity holds the relations

$$e_I(e_J e_K) = \theta_{IJK}(e_I e_J)e_K, \qquad (e_I e_J)e_K = \theta_{IJK} e_I(e_J e_K) \qquad (1.1.6)$$

From the tables 1.1.1a and 1.1.1c becomes apparent that the four first unit elements $1, e_x, e_y, e^z$ constitute a quaternion basis, while the rest four unit elements $e_z = ee^z,\ e^y = ee_y,\ e^x = e_x e,\ e = 1e$ are the unit elements of the quaternion basis left or right multiplied by the imaginary scalar unit $e$and constitute a *pseudo-quaternion* basis. Similar separation in quaternions and pseudo-quaternions can be applied to the totality of the rows and the columns of the table, each one representing an octonion basis referred to ident poising or counterpoising reference systems. Associativity is not valid for pseudo-quaternions. From this results a rule for the evaluation of the symbol $\theta_{IJK}$ in the relation (1.1.5). It is $\theta_{IJK} = -1$ for $IJK$denoting unit elements altogether belonging to a pseudo-quaternion basis, of those appearing in the semi-rows and the semi-columns of the table 1.1.1e, while otherwise it is $\theta_{IJK} = +1$.

### 1.1.1 Octonion Product

Given two octonions $a$ and $b$having the form defined by the relations (1.1.4) and (1.1.6)

$$a = \sum_{I=1}^{8} a_1 = a_1 + a_p + a^a + a^e \qquad\qquad b = \sum_{I=1}^{8} b_1 = b_1 + b_p + b^a + b^e$$

where by the capital letters $I,J$ are represented the eight low and upper right indices of the octonion components, the octonion product $ab$is equal to an octonion $C$ according to the relation

$$c = ab = \sum_{I=1}^{8} \sum_{J=1}^{8} a_1 b_J = c_1 + c_p + c^a + c^e \qquad (1.1.7)$$

where

## Table 1.1.1 - Multiplication Tables

| | 1 | $e_x$ | $e_y$ | $e^z$ | $e_z$ | $e^y$ | $e^x$ | $e$ |
|---|---|---|---|---|---|---|---|---|
| 1 | 1 | $e_x$ | $e_y$ | $e^z$ | $e_z$ | $e^y$ | $e^x$ | $e$ |
| $e_x$ | $e_x$ | -1 | $e^z$ | $-e_y$ | $e^y$ | $-e_z$ | $-e$ | $e^x$ |
| $e_y$ | $e_y$ | $-e^z$ | -1 | $e_x$ | $e^x$ | $e$ | $-e_z$ | $-e^y$ |
| $e^z$ | $e^z$ | $e_y$ | $-e_x$ | -1 | $e$ | $e^x$ | $e^y$ | $-e_z$ |
| $e_z$ | $e_z$ | $-e^y$ | $-e^x$ | $-e$ | -1 | $e_x$ | $e_y$ | $e^z$ |
| $e^y$ | $e^y$ | $e_z$ | $-e$ | $-e^x$ | $-e_x$ | -1 | $e^z$ | $e_y$ |
| $e^x$ | $e^x$ | $e$ | $e_z$ | $-e^y$ | $-e_y$ | $-e^z$ | -1 | $-e_x$ |
| $e$ | $e$ | $-e^x$ | $e^y$ | $e_z$ | $-e^z$ | $e^x$ | $e_x$ | -1 |

1.1.1a Sub-algebras arrangement
In counterpointing reference systems

| | 1 | $e_x$ | $e_y$ | $e^z$ | $e^x$ | $e^y$ | $e^z$ | $e$ |
|---|---|---|---|---|---|---|---|---|
| 1 | 1 | $e_x$ | $e_y$ | $e^z$ | $e^x$ | $e^y$ | $e^z$ | $e$ |
| $e_x$ | $e_x$ | -1 | $e^z$ | $e^y$ | $-e$ | $-e_z$ | $-e_y$ | $e^x$ |
| $e_y$ | $e_y$ | $-e^z$ | -1 | $e^x$ | $-e_z$ | $e$ | $e_x$ | $-e^y$ |
| $e_z$ | $e_z$ | $-e^y$ | $-e^x$ | -1 | $e_y$ | $e_x$ | $-e$ | $e^z$ |
| $e^x$ | $e^x$ | $e$ | $e_z$ | $e^x$ | -1 | $-e^z$ | $-e^y$ | $-e_x$ |
| $e^y$ | $e^y$ | $e_z$ | $-e$ | $-e_x$ | $e^z$ | -1 | $-e^x$ | $e_y$ |
| $e^z$ | $e^z$ | $e_y$ | $-e_x$ | $e$ | $e^y$ | $e^x$ | -1 | $-e_z$ |
| $e$ | $e$ | $-e^x$ | $e^y$ | $-e^z$ | $e_x$ | $-e_y$ | $e_z$ | -1 |

1.1.1b Scalars and vectors arrangement
In counterpointing reference systems

| | 1 | $e_x$ | $e_y$ | $e^z$ | $e_z$ | $e^y$ | $e^x$ | $e$ |
|---|---|---|---|---|---|---|---|---|
| 1 | 1 | $e_x$ | $e_y$ | $e^z$ | $e_z$ | $e^y$ | $e^x$ | $e$ |
| $e_x$ | $e_x$ | -1 | $e^z$ | $-e_y$ | $-e^y$ | $e_z$ | $-e$ | $e^x$ |
| $e_y$ | $e_y$ | $-e^z$ | -1 | $e_x$ | $e^x$ | $-e$ | $-e_z$ | $e^y$ |
| $e^z$ | $e^z$ | $e_y$ | $-e_x$ | -1 | $e$ | $e^x$ | $-e^y$ | $-e_z$ |
| $e_z$ | $e_z$ | $e^y$ | $-e^x$ | $-e$ | -1 | $-e_x$ | $e_y$ | $e^z$ |
| $e^y$ | $e^y$ | $-e_z$ | $e$ | $-e^x$ | $e_x$ | -1 | $e^z$ | $-e_y$ |
| $e^x$ | $e^x$ | $e$ | $e_z$ | $e^y$ | $-e_y$ | $-e^z$ | -1 | $-e_x$ |
| $e$ | $e$ | $-e^x$ | $-e^y$ | $e_z$ | $-e^z$ | $e_y$ | $e_x$ | -1 |

1.1.1c Sub-algebras arrangement
In counterpointing reference systems

| | 1 | $e_x$ | $e_y$ | $e_z$ | $e^x$ | $e^y$ | $e^z$ | $e$ |
|---|---|---|---|---|---|---|---|---|
| 1 | 1 | $e_x$ | $e_y$ | $e_z$ | $e^x$ | $e^y$ | $e^z$ | $e$ |
| $e_x$ | $e_x$ | -1 | $e^z$ | $-e^y$ | $-e$ | $e_z$ | $-e_y$ | $e^x$ |
| $e_y$ | $e_y$ | $-e^z$ | -1 | $e^x$ | $-e_z$ | $-e$ | $e_x$ | $e^y$ |
| $e_z$ | $e_z$ | $e^y$ | $-e^x$ | -1 | $e_y$ | $-e_x$ | $-e$ | $e^z$ |
| $e^x$ | $e^x$ | $e$ | $e_z$ | $-e_y$ | -1 | $-e^z$ | $e^y$ | $-e_x$ |
| $e^y$ | $e^y$ | $-e_z$ | $e$ | $e_x$ | $e^z$ | -1 | $-e^x$ | $-e_y$ |
| $e^z$ | $e^z$ | $e_y$ | $-e_x$ | $e$ | $-e^y$ | $e^x$ | -1 | $-e_z$ |
| $e$ | $e$ | $-e^x$ | $-e^y$ | $-e^z$ | $e_x$ | $e_y$ | $e_z$ | -1 |

1.1.1d Scalars and vectors arrangement
In counterpointing reference systems

$$c_1 = (ab)_1 = a_1 b_1 + a_p \cdot b_p + a^a \cdot b^a + a^e b^e$$

$$c_p = (ab)_p = a_1 b_p + a_p b_1 + a_p \otimes b^a + a^a \otimes b_p + a^a b^e + a^e b^a$$

$$c^a = (ab)^a = a_1 b^a + a^a b_1 + a_p \times b_p + a^a \times b^a + a_p b^e + a^e b_p \qquad (1.1.7a)$$

$$c_e = (ab)_e = a_1 b^e + a^e b_1 + a_p \circ b^a + a^a \circ b_p$$

and

$$a_p \cdot h_p = \sum_i a_i b_i, \quad a^a \cdot b^a = \sum_i a^i b^i$$

The commutative *dot products* of polar by polar and axial by axial vectors equivalent to real scalars,

$$a_p \circ b^a = \sum_i a_i b^i, \quad a^a \circ b_p = \sum_i a^i b_i$$

The non-commutative *hole products* of polar by axial and axial by polar vectors equivalent to imaginary scalars,

$$a_p \times b_p = \sum_i \sum_{j \neq i} a_i b_j, \quad a^a \times b^a = \sum_i \sum_{j \neq i} a^i b^j$$

The non-commutative *cross products* of polar by polar and axial by axial vectors equivalent to real axial vectors,

$$a_p \otimes b^a = \sum_i \sum_{j \neq i} a_i b^j, \quad a^a \otimes b_p = \sum_i \sum_{j \neq i} a^i b_j$$

The non-commutative *grid products* of polar by axial and axial by polar vectors equivalent to imaginary polar vectors

Where by the prosaic letters $i,j$ are denoted the three indices $x, y, z$ independenty of their upper or low right position Eqs. (1.1.7a) can be represented by the following more compact and mindful form:

$$c_1 = a \cdot b \quad = \begin{bmatrix} a_1 & & a_p & & a^a & & a_e \\ \downarrow & \times & \downarrow & \times & \downarrow & \times & \downarrow \\ b_1 & & b_p & & b^a & & b^e \end{bmatrix}$$

$$c_p = a \otimes \vec{b} = \begin{bmatrix} a^1 & & a_p & & a^a & & a_e \\ & \times & & \times & & \times & \\ b^1 & & b_p & & b^a & & b_e \end{bmatrix}$$

(1.1.7b)

$$c^a = a \times \overleftarrow{b} \;= \begin{bmatrix} a^1 & a_p & a^a & a_e \\ & & & \\ b_e & b^a & b_p & b^1 \end{bmatrix}$$

$$c_e = a \circ \overleftarrow{b} \quad = \begin{bmatrix} a^1 & a_p & a^a & a_e \\ \downarrow & \downarrow & \downarrow & \downarrow \\ b_e & b^a & b_p & b^1 \end{bmatrix}$$

The right directed overhead arrow $\rightarrow$indicates that the terms of $b$are taken in the normal order, while the left directed arrow $\leftarrow$ indicates that the terms of $b$ are taken in the inverted order. The expressions in [ ] represent the sum of the products indicated by the downward directed arrows, which are of the type defined by the symbol between $a$ and $b$ in the abridged expression. More briefly, for parallel arrows the products are of the dot or hole type, while for crossing arrows the products are of the cross or grid type.

Due to the non-associativity the product of more than two octonions have to be analyzed in two – term products through the appropriate insertion of parentheses showing the sequence of the successive operations. For example, from the expression $\big(a(bc)\big)d$ results that initially is executed the operation $bc$ secondly the $a(bc)$ and finally the$\big(a(bc)\big)d$. Concerning the terms of the product containing the corresponding octonion components it is purposeful at the right of the parenthesis including a partial product to be written the index corresponding to the indices of the two factors of the two-term product according to the multiplication table 1.1.1c, d, so that each two – term product to be considered as a single factor of the two – term product contained in the next parenthesis. After repeated application of this procedure will result finally the index of the component corresponding to the considered term of the product. For example the term containing the components $\left(a_y\left(b^x c_y\right)\right) d^z$ is written $\left(\left(a_y\left(b^x c_y\right)_z\right)^x d^z\right)^y$.

More analytic expressions for the cross and grid products are the following:

$$\begin{aligned}
\left(a_p \times b_p\right)^a &= \sum\left(a_i b_j + a_j b_i\right)^k = \sum a_i\left(b_j + b_k\right) \\
&= \sum\left(a_i + a_j\right) b_k \\
\left(a^a \times b^a\right)^a &= \sum\left(a^i b^j + a^j b^i\right)^k = \sum a^i\left(b^j + b^k\right) \\
&= \Sigma\left(a^i + a^j\right) b^k
\end{aligned}$$

$$\begin{aligned}
\left(a_p \otimes b^a\right)_p &= \sum\left(a_i b^j + a_j b^i\right)_k = \sum a_i\left(b^j + b^k\right) \\
&= \sum\left(a_i + a_j\right) b^k \\
\left(a^a \otimes b_p\right)_p &= \Sigma\left(a^i b_j + a^j b_i\right)_k = \Sigma\, a^i\left(b_j + b_k\right) \\
&= \sum\left(a^i + a^j\right) b_k
\end{aligned} \tag{1.1.8}$$

with $i, j, k$ denoting the three cyclic permutations of $x, y, z$ contained in the sum.

The first expression for the cross product $a_p \times b_p$ written in terms of scalar coefficients and unit vectors, has the form, after taking into account that $e_k e_j = -e_j e_k = -e^i$

$$\left(a_p \times b_p\right)^a = \sum\left(\hat{a}_j \hat{b}_k - \hat{a}_k \hat{b}_j\right)\left(e_j e_k\right)^i = \sum\left(\hat{a}_j \hat{b}_k - \hat{a}_j \hat{b}_j\right) e^i \tag{1.1.8a}$$

which coincides with that defined for the ordinary vector representation. Moreover the cross product can be represented by the determinant form

$$\left(a_p \times n_p\right)^a = \begin{vmatrix} 1 & 1 & 1 \\ a_x & a_y & a_z \\ b_x & b_y & b_z \end{vmatrix} \qquad (1.1.8b)$$

Similar to (1.1.8b) expressions are applied for the rest cross and wheel products after setting the indices of the components in the appropriate upper or low right position.

The octonion product *ab* can furthermore be split in two terms

$$ab = a\Lambda b + aVn \qquad (1.1.9)$$

where

$$\begin{aligned} a\Lambda b &= \frac{1}{2}(ab + ba) \\ &= \left(a_1 b_1 + a_p \cdot b_p + a^a \cdot b^a + a^e b^e\right)_1 + \left(a_1 b_p + a_p b^1\right)_p + \\ &\quad (a_1 b^a + a^a b_1)^a + (a_1 b^e + a^e b^e)^e \end{aligned} \qquad (1.1.9a)$$

is the *commutator* of $a$ and $b$, equal to the sum of the commutative terms of the product and

$$\begin{aligned} aVb &= \tfrac{1}{2}(ab - ba) = \left(a_p \otimes ba + a^a \otimes b_p + a^a b^e + a^e b^a\right)_p + \\ &\quad \left(a_p \times b_p + a^a \times b^a + a_p b^e + a^e b_p\right)^a \end{aligned}$$

$$(1.1.9b)$$

is the *anti-commutator* of $a$ and $b$, equal to the sum of the non-commutative terms of the product.

In the special cases where the octonions consist exclusively of a single polar or a single axial vector the octonion product has the form

$$a_p b_p = \left(a_p \cdot b_p\right)_1 + \left(a_p \times b_p\right)^a \qquad (1.1.10a)$$

$$a^a b^a = (a^a \cdot b^a)_1 + (a^a \times b^a)^a \qquad (1.1.10b)$$

$$a_p b^a = \left(a_p \circ b^a\right)^e + \left(a_p \otimes b^a\right)_p \qquad (1.1.10c)$$

$$a^a b_p = \left(a^a \circ b_p\right)^e + \left(a^a \otimes b_p\right)_p \qquad (1.1.10d)$$

and

$$a_p^2 = \sum_i a_i^2 = -\sum \hat{a}_1^2, \qquad a^{a2}$$
$$= \sum_i a^{i2} = -\sum \hat{a}^{i2} \qquad (1.1.11)$$

The following relations

$$a_p \cdot b_p = 0, \quad a^a \cdot b^a = 0, \quad a_p \circ b^a = 0, \quad a^a \circ b_p = 0 \qquad (1.1.12)$$

constitute the *orthogonality conditions* between vectors, while the relations

$$a_p \times b_p = 0, \quad a^a \times b^a = 0, \quad a_p \otimes b^a = 0, \quad a^a \otimes b_p = 0 \qquad (1.1.13)$$

constitute the *collinearity conditions* between vectors.

### 1.1.2 Conjugates, Norms, Inverses, Reverses

The *conjugate* $\tilde{a}$ of an octonion $a = a_1 + a_p + a^a + a^e$ is the octonion

$$\tilde{a} = a_1 - a_p - a^a - a^e = a_1 - a_x - a_y - a_z - a^x - a^y - a^z - a^e \qquad (1.1.14)$$

The product $a\tilde{a}$ is the norm $N[a] = N[\tilde{a}]$ of the octonions $a$ and $\tilde{a}$. According to the relations (1.1.9, 10, 11) it is

$$N[a] = N[\tilde{a}] = a\tilde{a} = a\Lambda\tilde{a} = \sum_{I=1}^{8} \hat{a}_I^2 = |a|^2 = |\tilde{a}|^2 \qquad (1.1.15)$$

and the norm is a non-negative real scalar. The non-negative real scalar $|a| = N[a]^{\frac{1}{2}}$ is the *length* of the octonion $a$. It is easily verified that

$$(\widetilde{ab}) = \tilde{b}\tilde{a} \tag{1.1.16}$$

and

$$N[ab] = N[a] \cdot N[b] \tag{1.1.17}$$

From (1.1.18) results that the octonion algebra has no divisors of zero. The octonion

$$\hat{a} = \sum_{I=1}^{8} \hat{a}_I = \frac{\tilde{a}}{|a|^2} \tag{1.1.18}$$

is the *inverse* of $a$. According to (1.2.16) the following relation is fulfilled:

$$a\hat{a} = \hat{a}a = \frac{a\hat{a}}{|a|^2} = 1 \tag{1.1.19}$$

Another characteristic octonion associated with any given octonion $a$, is the *reverse* octonion $\bar{a}$ . The components $\bar{a}_I$ of $\bar{a}$ are the inverses of the components $a_I$ of $a$, each one considered as a single octonion. It is

$$a = \sum_{I=1}^{8} \bar{a}_I = \sum_{I=1}^{8} \frac{a_I}{a_I^2}, \qquad a_I\bar{a}_I = 1 \tag{1.1.20}$$

The product $a\bar{a}$ is a positive integer equal to the number of the non-zero components of $a$.

For the unit elements $e_I$ of the octonion basis, in case each one is considered as a single octonion with $|e_I|^2 = 1$, the following relations are fulfilled:

$$\tilde{e}_1 = \hat{e}_1 = \bar{e}_1 = e_1 = 1, \quad \{I = 1\} \tag{1.1.21a}$$
$$\tilde{e}_1 = \hat{e}_1 = \bar{e}_1 = -e_1, \quad \{I \neq 1\} \tag{1.1.21b}$$
$$e_I\tilde{e}_I = e_I\hat{e}_I = e_I\bar{e}_I = |e_I{}^2| = 1 \tag{1.1.21c}$$

On the bases of the above relations the multiplication tables 1.1.1c,d,e,f receive the form of the multiplication tables 1.1.2a,b,c,d, where the unit elements $-e_I$ for $I \neq 1$, are replaced by the equivalent $\bar{e}_I$ which represent likewise and the symbols $\tilde{e}_I$ and $\hat{e}_I$. For uniformity purposes of the notation is likewise introduced $-1 = \bar{1}$ taking into consideration that contrary to the overall accepted notation, the symbol $\bar{1}$ does not represent any one of the symbols $\tilde{e}_1, \hat{e}_1, and\ \bar{e}_1$ all them being equal to 1. The superscript dash of the element resulting from the multiplication table, is also put on the index at the right of each parenthesis including two-term products which constitute a multi-term product component. For example the product component $\left(a_y\left(b_y c^x\right)\right) d^z$ is written $\left(\left(a_y\left(b_y c^x\right)_{\bar{z}}\right)^{\bar{x}} d^z\right)^{\bar{y}}$. In this procedure the relations (1.1.22) have to be taken into account.

The minus sign before a single or a multi-product octonion component is equivalent to the dash over the index, as both specify that the direction of the corresponding unit element is opposite to the normal one. In the case of an octonion equation expressed by a sum of terms which are referred to the same unit element, the transfer of terms from the one side of the equation to the other, by the simple change of their sign, is not permitted, because by this change the reference unit element is altered. For example, the equation $a_I - b_{\bar{I}} = 0$ does not imply $a_I = b_{\bar{I}}$, as $I \neq \bar{I}$. The transfer of terms from one side to the other becomes possible by changing only the sign of the numerical value of the component and this will be denoted by a dash under the symbol of the term transferred. In the above example the equation after the transfer will have the form $a_I - b_{\bar{I}}$, with $I = -\bar{I}$.

Finally, it is pointed out that the non-commutativity relation (1.1.3) and the non-associativity – alternativity relations (1.1.6) are unaltered under the presence of superscript dashes over the elements $e_I$ and are dependent exclusively from the indices $I, J, K$.

### 1.1.1 Octonion Tensors

An *octonion space* ${}^{a}_{n}\omega$ is a set of octonions ${}_{i}a$, referred to an octonion basis consisting of $n$ unit octonion elements $a_{e_I}$, which represent geometrical or natural entities of the same kind. Examples of the different kinds of entities are the length, the area, the displacement, the velocity, the force, the momentum, etc. The octonions, compared to the ordinary vectors, possess the additional property to be elements of a division algebra, this consisting the main difference between octonion spaces and ordinary vector spaces.

**Table 1.1.2 - Multiplication Tables (in reverse terms)**

| | $1$ | $e_x$ | $e_y$ | $e^z$ | $e_z$ | $e^y$ | $e^x$ | $e$ |
|---|---|---|---|---|---|---|---|---|
| $1$ | $1$ | $e_x$ | $e_y$ | $e^z$ | $e_z$ | $e^y$ | $e^x$ | $e$ |
| $e_x$ | $e_x$ | $\bar{1}$ | $e^z$ | $\bar{e}_y$ | $\bar{e}^y$ | $e_z$ | $e$ | $e^x$ |
| $e_y$ | $e_y$ | $\bar{e}^z$ | $\bar{1}$ | $e_x$ | $e^x$ | $\bar{e}$ | $\bar{e}_z$ | $e^y$ |
| $e^z$ | $e^z$ | $e_y$ | $\bar{e}_x$ | $\bar{1}$ | $e$ | $e^x$ | $\bar{e}^y$ | $\bar{e}_z$ |
| $e_z$ | $e_z$ | $e^y$ | $\bar{e}^x$ | $\bar{e}$ | $\bar{1}$ | $\bar{e}_x$ | $e_y$ | $e^z$ |
| $e^y$ | $e^y$ | $\bar{e}_z$ | $e$ | $\bar{e}^x$ | $e_x$ | $\bar{1}$ | $e^z$ | $\bar{e}_y$ |
| $e^x$ | $e^x$ | $e$ | $e_z$ | $e^y$ | $\bar{e}_y$ | $\bar{e}^z$ | $\bar{1}$ | $\bar{e}_x$ |
| $e$ | $e$ | $\bar{e}^x$ | $\bar{e}^y$ | $e_z$ | $\bar{e}^z$ | $e_y$ | $e_x$ | $\bar{1}$ |

1.1.2a Sub-algebras arrangement

| | $1$ | $e_x$ | $e_y$ | $e_z$ | $e^x$ | $e^y$ | $e^z$ | $e$ |
|---|---|---|---|---|---|---|---|---|
| $1$ | $1$ | $e_x$ | $e_y$ | $e_z$ | $e^x$ | $e^y$ | $e^z$ | $e$ |
| $e_x$ | $e_x$ | $\bar{1}$ | $e^z$ | $\bar{e}^y$ | $\bar{e}$ | $e_z$ | $\bar{e}_y$ | $e^x$ |
| $e_y$ | $e_y$ | $\bar{e}^z$ | $\bar{1}$ | $e^x$ | $\bar{e}_z$ | $\bar{e}$ | $e_x$ | $e^y$ |
| $e_z$ | $e_z$ | $e^y$ | $\bar{e}^x$ | $\bar{1}$ | $e_y$ | $\bar{e}_x$ | $\bar{e}$ | $e^z$ |
| $e^x$ | $e^x$ | $e$ | $e_z$ | $\bar{e}_y$ | $\bar{1}$ | $\bar{e}^z$ | $e^y$ | $\bar{e}_x$ |
| $e^y$ | $e^y$ | $\bar{e}_z$ | $e$ | $e_x$ | $e^z$ | $\bar{1}$ | $\bar{e}^x$ | $\bar{e}_y$ |
| $e^z$ | $e^z$ | $e_y$ | $\bar{e}_x$ | $e$ | $\bar{e}^y$ | $e^x$ | $\bar{1}$ | $\bar{e}_z$ |
| $e$ | $e$ | $\bar{e}^x$ | $\bar{e}^y$ | $\bar{e}^z$ | $e_x$ | $e_y$ | $e_z$ | $\bar{1}$ |

1.1.2b Scalars and vectors arrangement

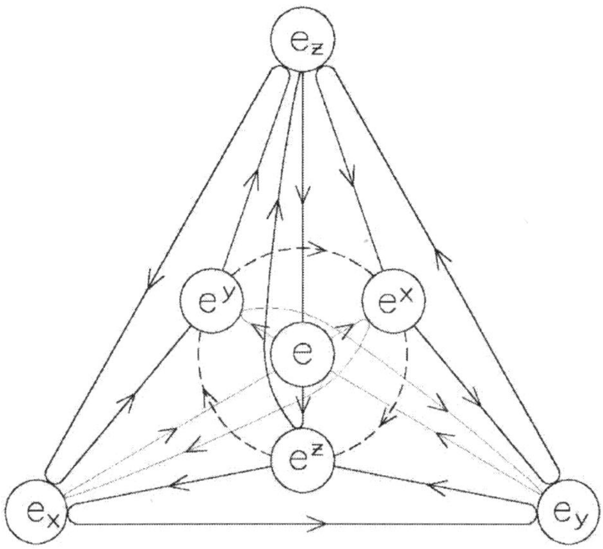


*Figure 1.1.2 - Fano Plane*

Concerning the herefore introduced notation, left upper and low indices are used to specify different octonion spaces and sets for the enumeration of the contained elements, but never as a component notation, this provided exclusively by the right upper and low indices. For that reason the left indices don't contribute and have no effect to the operations between octonions.

An *octonion tensor space* ${}^{ab}_{nm}\omega$ of dimensionality $nm$ is the tensor product of two octonion spaces ${}^{a}_{n}\omega$ and ${}^{b}_{m}\omega$ consisting of octonions ${}_{i}a$ and ${}_{j}b$ referred to bases of unit elements $a_{e_I}$ and $b_{e_I}$ respectively, and contains the set of all the octonion products ${}^{ab}_{ij}c = {}_{i}a\ {}_{j}b$ called octonion tensors, consisting of $nm$ octonion components ${}^{ab}_{ij}c_{IJ} = {}_{i}a_{I}|\ {}_{i}b_{J}$ refereed to a basis of $nm$ unit elements ${}^{a}e_{I}\ {}^{b}e_{J} = {}^{ab}e_{K}$ which satisfy the following conditions:

$$(\lambda a)b = a(\lambda b) = \lambda(ab), \qquad \{\lambda\!: real\ scalar\} \tag{1.1.22a}$$

$$a(\ {}_{i}b +\ {}_{j}b = a|\ {}_{i}b + a|\ {}_{j}b) \tag{1.1.22b}$$

$$(\ {}_{i}a +\ {}_{j}a)b =\ {}_{i}ab +\ {}_{j}ab \tag{1.1.22c}$$

The vertical bar will be used any time is needed to separate the left indices of a symbol from the right indices of the previous one and does not affect the operation between them.

The conditions (1.1.22) are obviously satisfied in the case of parallel bases of the two octonion spaces. Because of the orthogonality of the octonion bases there is no distinction between covariant and contravariant octonion components, represented in the ordinary tensor algebra by right low and upper indices, which in the present notation symbolize exclusively the and octonion components respectively.

A tensor component, being equivalent to a two-term product of two octonion components, and contributing to a multi-term product, is purposeful, following the already introduced notation, to be included in a parenthesis with an external right index corresponding to the indices of the two factors according to the multiplication table. For example the tensor component corresponding to the octonion product $(a_x b_y)^z$ is written $(c_{xy})^z$. Moreover the basis of an octonion tensor, independently of its dimensionality $nm$, is composed of unit elements $\left( {}^a e_I \middle| {}^a e_J \right)_K$ equivalent to unit elements belonging to the set of the 8 elements $e_K$ of the octonion basis and the set of the 8 inverse elements $\bar{e}_{K,} K$ being the index corresponding to the indices $I$ and $J$, according to the multiplication table. The distinction between tensor unit elements equivalent to the same octonion unit element results from the different sets of the three indices. In case of formulae containing products of octonion and tensor components is applicable the Einstein's summation convention permitting the summation sign to be suppressed in summations effected with respect to an index $I$ repeated twice in the product.

The product of two complete octonions belonging to two different octonion spaces is a tensor consisting of 64 components. Other tensors with smaller dimensionality contributing in the present treatise are the following:

The *strain tensor* with 3X3=9 components $(\varepsilon_{\bar{\iota}\bar{J}})^{\bar{k}}$, $(\varepsilon_{i\bar{\iota}})_1$,

The *stress tensor* with 3X3=9 components $(\sigma_i{}^{\bar{J}})_{\bar{k}}$, $(\sigma_i{}^{\bar{\iota}})^e$

The two *shape function* tensors, a conformal ${}_c\zeta_{\bar{I}i}$ and a contra-formal ${}^c\zeta_{\bar{I}i}$ each one with 8X3=24 components, equal to the corresponding to them *lever-arm function* tensors ${}_c\xi_{I\bar{\iota}}$ *and* ${}^c\xi_{I\bar{\iota}}$

where,

$i, j, k$ : are permutations of the three indices $x, y, z$

$I$ : is anyone of the eight indices of the octonion components

### 1.1.2 Polar and Axial Vectors

In this section are presented, in the octonion notation introduced in the previous sections, properties of polar and axial vectors already known from the ordinary algebra of vector spaces. According to the relations (1.1.20, 22b and 12a) a polar vector $b_p = b_x + b_y + b_z = \sum b_i$ satisfies the relation

$$b_p \sum (\bar{e}_i e_i)_1 = 3b_p = \sum_i (b_p \bar{e}_i) e_i = \sum_i (b_p \cdot \bar{e}_i)\, e_i + \sum_i (b_p \times \bar{e}_i)\, e_i =$$
$$= \sum_i (b_p \cdot \bar{e}_i) e_i + 2b_p$$

because, according to the relation (1.1.8), it is

$$\sum_i (b_p \times \bar{e}_i) e_i = \sum_i \big((b_j + b_k)\bar{e}_i\big)\, e_i$$
$$= \sum_i (b_j + b_k)(\bar{e}_i e_i)_1 = 2\big(b_x + b_y + b_z\big) = 2b_p$$

From the above relations results

$$b_p = \sum (b_p \cdot \bar{e}_i)_1 e_i = \sum b_{p\cdot\bar{\imath}}\, e_i = \sum b_i \qquad (1.1.23)$$

where, $b_{p\cdot\bar{\imath}} = (b_p \cdot \bar{e}_i)_1$ is the real scalar coefficient of $e_i$ of the component $b_i$, already introduced in section 1.1.1, equal to the projection of $b_i$ on the unit vector $e_i$. Moreover with any polar vector $b_p$ are associated, the norm:

$$|b_p|^2 = -(b_i b_i)_1 = -\sum_i b_i^2 = -b_p^2 \qquad (1.1.24a)$$

the inverse vector:

$$\hat{b}_p = -\frac{1}{|b_p|^2} b_p, \quad \hat{b}_i = -\frac{1}{|b_p|^2} b_i, \quad |\hat{b}_p|^2 = -\frac{1}{|b_p|^2}, \quad b_p \hat{b}_p = 1 \qquad (1.1.24b)$$

and the unit vector, collinear with $b_p$,

$$e_p = \frac{1}{|b_p|} b_p = \sum_i e_{b\cdot\bar{\imath}} e_i, \qquad e_b^2 = \frac{1}{|b_p|^2} b_p^2 = -1 \qquad (1.1.24c)$$

where,

$$e_{b\cdot\bar{\imath}} = (e_b \cdot \bar{e}_i)_1 = \frac{1}{|b_p|} (b_p \cdot \bar{e}_i)_1 = \frac{b_{p\cdot\bar{\imath}}}{|b_p|}$$

are the *direction cosines* of the vector $b_p$ with respect to the $i$ axes, equal to the projections of the components of $e_b$ on these axes.

Relations similar to the above are applied for axial vectors $b^a = b^x + b^y + b^z = \sum b^i$ by changing the indices from the low to the upper position and the index $p$ by the index $a$.

### 1.1.3 Orthogonal Transformations

The scalar coefficients of the components of the unit vectors $e_{\bar{i}}$ of a second polar vector orthogonal basis, resulting from the rotation of the original basis around a random axis, are equal to their direction cosines with respect to the unit vectors $e_i$ of the original basis, according to the relation

$$e_{\bar{i}'} = \sum_j e_{\bar{i}\cdot\bar{j}} e_j \tag{1.1.25}$$

which defines an *Orthogonal Transformation*, from the basis $e_{\bar{i}}$ to the basis $e_{\bar{i}'}$. The nine direction cosines $e_{\bar{i}\cdot\bar{j}}$ satisfy the following three orthogonality conditions

$$-(e_{\bar{i}'} \cdot e_{\bar{j}'}) = \sum_k e_{\bar{i}'\bar{k}}\, e_{\bar{j}'\bar{k}} = 0 \tag{1.1.26a}$$

as well as the three unit norm conditions

$$|e_{\bar{i}'}|^2 = -e_{\bar{i}'}^2 = \sum_j e_{\bar{i}'\bar{j}}^2 = 1 \tag{1.1.26b}$$

The unit vectors $e_{\bar{i}}$ of the original basis are presented in the derived basis of the unit vectors $e_{\bar{i}'}$ through the following relation, similar to the above (1.1.26)

$$e_j = \sum_{j'} e_{j\cdot\bar{i}'} - e_{\bar{i}'} \tag{1.1.27}$$

with $e_{j\cdot\bar{i}'}$ satisfying the conditions, similar to (1.1.27),

$$\sum_{k} e_{i\cdot\bar{k}} e_{j\cdot\bar{k}} = 0, \qquad \sum_{i'} e_{j\cdot i'}^{2} = 1 \qquad (1.1.28)$$

It is easily verified that $e_{i\cdot\acute{\bar{j}}} = \left(e_{i'} \cdot \bar{e}_j\right)_1 = \left(e_j \cdot \bar{e}_{i'}\right)_1 = e_{j\cdot\bar{\imath}}$ (1.1.29)

The equations (1.1.26, 28 and 30) are equivalent to the matrix relations

$$\lfloor e_{i\cdot\acute{\bar{j}}} \rfloor = \lfloor e_{j\cdot\bar{\imath}'} \rfloor_{\bar{1}} = \lfloor e_{j\cdot\bar{\imath}'} \rfloor_1 \qquad (1.1.30)$$

and according to a known theorem of the algebra of matrices, the corresponding determinants have the value

$$det\lfloor e_{i\cdot\acute{\bar{j}}} \rfloor = det\lfloor e_{j\cdot\bar{\imath}'} \rfloor = 1 \qquad (1.1.31)$$

The elements $e_{i\cdot\acute{\bar{j}}}$ and $e_{j\cdot\bar{\imath}'}$ of the transformation matrices don't satisfy the conditions (1.1.23) and hence they are not elements of tensors.

The additional three conditions required for the evaluation of $e_{i\cdot\acute{\bar{j}}}$ are their relations with three parameters describing the rotations of the original basis through which the second basis is obtained. The following rotation procedures defining the required three parameters are offered:

- Two direction cosines of the axis of rotation defined by the unit vector

$$e^{\phi} = e^{\phi\cdot\bar{\imath}} e^{i} + e^{\phi\cdot\bar{\jmath}} e^{j} + e^{\phi\cdot\bar{k}} e^{k}, \qquad \left|e^{\phi}\right|^2 = \sum \left(e^{\phi\bar{\imath}}\right)^2 = 1 \qquad (1.1.32)$$

and the angle of rotation $\hat{\phi}$ around it. By these set of parameters is defined the Hamilton's quaternion

$$q = cos\frac{\hat{\Phi}}{2} + e^{\Phi}\, sin\frac{\hat{\Phi}}{2} \qquad (1.1.33)$$

The transformed $b'$ of a vector $b$, resulting after the rotation $\phi = |\phi| e^{\Phi}$ is

$$b' = qb\bar{q} \qquad (1.1.34)$$

The products with quaternions has to be executed in the counterpointing system for $e^x, e^y, e^z$ where the following relations have to be taken into account

$$e_z e_x = \bar{e}^y, \leftrightarrow e^z e_x = (e\bar{e}_z)^z e_x = -(e(\bar{e}_z e_x)^y)_y = \bar{e}_y \tag{1.1.34a}$$

instead of those resulting in the identpointing system $e^y$ and $e_y$ respectively.

- The three Euler's angles $\phi, \psi, \theta$ around three constant axes, already contained in detail in the existing literature on the subject.
- The three successive rotation angles $\phi^x, \phi^{y'}, \phi^{z''}$around the transformed axes of the original basis for its three positions resulting successively for each one of the above rotations. (Fig. 1.1.3)

The components $b_{i'}$ in the transformed basis, of a polar vector $b_p$ with components $b_i$ in the original basis, result after the substitution of the relations (1.1.26, 30) in the (1.1.24), according to the relation

$$b_p = \sum_i b_{p\cdot\bar{i}} e_i = \sum_i b_{p\cdot\bar{i}} \left( \sum_{j'} e_{i\cdot\bar{j}'} e_{j'} \right) = \sum_{j'} \left( \sum_i b_{p\cdot\bar{i}} e_{i\cdot\bar{j}} \right) e_{j'} =$$

$$= \sum_{j'} b_{p\cdot\bar{j}'} e_{j'} = \sum_{j'} b_{j'}$$

and

$$b_{j'} = b_{p\cdot\bar{j}} e_{j'} \tag{1.1.35}$$

Because of the relations the following scalar entities remain invariant under an orthogonal transformation.

The norm of a polar vector $b$

$$|b'|^2 = \sum b\frac{2}{\bar{i}'} = \sum b_j b_k e_{j\cdot i'} - e_{k\cdot i'} = \sum b_j^2 e_{j\cdot i'}^2 = \sum b_j^2 = |b|^2$$

The norm of the inverse polar vector

$$|\hat{b}|^2 = \frac{1}{|b|^2}$$

The dot product of two polar vectors $a_p$ and $b_p$

$$-a' \cdot b' = a_{i'}b_{i'} = a_j b_j e_{j\cdot i'} - e_{k\cdot j'} = a_j b_j e^2_{j\cdot i'} = a_j b_j = -a \cdot b$$

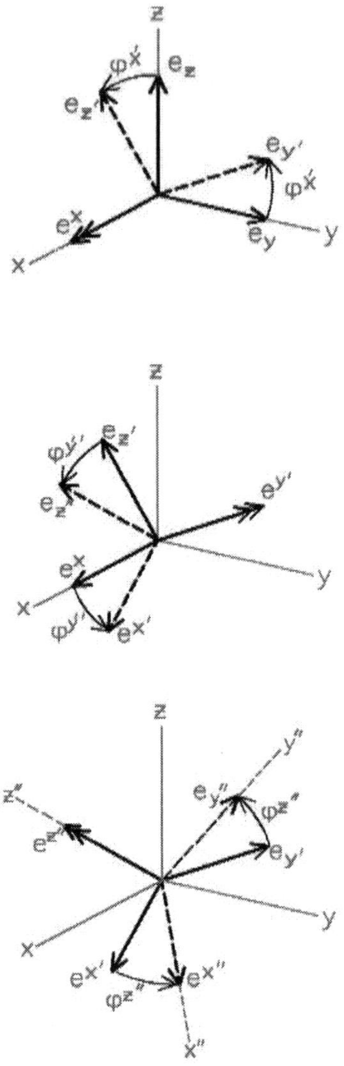


*Figure 1.1.3 - Rotation Angles*

The angle $\theta$ between two polar vectors $a$ and $b$

$$cos\,\theta = -\frac{a \cdot b}{|a||b|}$$

The norm of the cross product of two polar vectors $a_p$ and $b_p$

$$|a \times b| = |a||b|sin\theta$$

In a similar manner is easily verified that the following scalar quantities, concerning axial vectors, remain invariant under orthogonal transformations:

$$|b^a|^2, \quad |\hat{b}^a|^2, \quad a^a \cdot b^a, \quad |a^a \times b^a|, \quad a^a \circ b_p, \quad |a^a \otimes b_p|$$

as well as the angles between any two vectors polar or axial.

### 1.1.4 Position Vectors

To an Euclidean point space defined in an orthogonal system of $x, y, z$ coordinates having origin at a point 0, is associated a polar vector space so that, at any point A having coordinates ${}_Ax,\ {}_Ay,\ {}_Az$ corresponds a *position vector.*

$${}_Ar = {}_Axe + {}_Aye + {}_Aze_z = {}_Ar_x + {}_Ar_y + {}_Ar_z$$

Especially for the position vectors, the coordinates will be considered $x, y, z$ in the sequence as vector components in the respective directions instead of the $r_x, r_y, r_z$ and not as real scalar coefficients, so that a position vector receives the form

$$r = x + y + z = \sum_i r_i \tag{1.1.36a}$$

to which are associated the norm

$$|r|^2 = -r^2 = -(x^2 + y^2 + z^2) = -\sum_i r_i^2 \tag{1.1.36b}$$

and the unit vector

$$e_r = \frac{r}{|r|} = \frac{1}{r}(x + y + z) = \frac{1}{|r|}\sum_i r_i \qquad (1.1.36c)$$

with direction cosines

$$e_{r\bar{\iota}} = (e_r \cdot \bar{e}_\iota)_1 = \frac{1}{|r|}(r \cdot \bar{e}_i)_1 = \frac{1}{|r|}(r_i \bar{e}_i)_1 \qquad (1.1.36d)$$

If ${}_A r$, ${}_B r$ are the position vectors of two points $A$ and $B$ respectively, the vector is the *translation vector* from the point $A$ to the point $B$.

$$ {}_{AB}u = {}_B r - {}_A r, \qquad {}_B r = {}_A r + {}_{AB}u$$

With a position vector $r$ are furthermore associated:

The inverse position vector $\hat{r}$ which satisfies the relation

$$\hat{r} = \sum r_i = x_i + y_i + z_i, \qquad (1.1.37a)$$

$$r_i = -\frac{r_i}{|r|^2} = \frac{r_i}{|r_i|^2}\frac{|r_i|^2}{|r|^2} = \hat{r}_i e_{r\bar{\iota}}^2$$

The reverse position vector $\bar{r}$, non collinear with $r$,

$$\bar{r} = \sum_i \bar{r}_i = \bar{x}, \bar{y}, \bar{z}, \qquad \bar{r}_i = -\frac{r_i}{|r_i|^2}, \qquad r_i \bar{r}_i = 1 \qquad (1.1.37b)$$

The axial vector of the *area coordinates*

$$r^a = (yz)^x + (zy)^y + (xy)^z \qquad (1.1.37c)$$

The imaginary volume scalar

$$r^e = ((yz)^x x)^e = ((zx)^y y)^e = ((xy)^z z)^e \qquad (1.1.37d)$$

The vectors $\bar{r}, r^a$ and $r^e$ satisfy the relation implying the collinearity of $r^a$ and $\bar{r}$.

$$r^a = r^e \bar{r} \quad (1.1.38)$$

The scalar entities $|\bar{r}|^2, |r^a|^2$ and $|r^e|^2$ are not invariant under an orthogonal transformation. For one at least component $r_i = 0$ of $r$ they receive the values $|\bar{r}|^2 = \infty$ and $|r^a|^2 = |r^e|^2 = 0$. For one of the eight equiangular positions of $r$ with respect to the three coordinate axes corresponding to components $|x| = |y| = |z| = \frac{|r|}{\sqrt{3}}$ the three vectors $r, \bar{r}$ and $r^a$ are collinear, with norms receiving the following extreme values:

$$min|\bar{r}|^2 = \frac{3}{|r|^2} = 3|\bar{r}|^2 \quad (1.1.39a)$$

$$max|r^a|^2 = \frac{|r|^4}{3} \quad (1.1.39b)$$

$$max|r_e|^2 = \frac{|r|^6}{27} \quad (1.1.39c)$$

### 1.1.8 Free Polar Vectors

A polar vector $b$ applied at a point of the Euclidean point space corresponding to the position vector $r$ is analyzed to two characteristic components according to the following procedure, already applied in the derivation of the relation (1.1.24):

$$b = b(\bar{e}_r e_r)_1 = (b\bar{e}_r)e_r = ((b \cdot \bar{e}_r)_1 e_r)_r + ((b \times \bar{e}_r)^s e_r)_t = b_r + b_t \quad (1.1.40)$$

The first component $b_r = ((b \cdot \bar{e}_r)_1 e_r)_r$ is the projection of $b$ in the direction of $e_r$, having a scalar value $(b \cdot \bar{e}_r)_1$, and constitutes the radial component of $b$. The cross product

$$b^s = (e_r \times b)^s = \frac{1}{|r|}(r \times b)^s \quad (1.1.40a)$$

is an axial vector orthogonal to both $b$ and $r$ (Fig 1.2.4). The second component

$$b_t = ((b \times \bar{e}_r)^s e_r)_t = (b^s e_r)_t = (b^s \otimes e_r)_t \quad (1.1.40b)$$

It is a polar vector orthogonal to both $b^s$ and $e_r$ and constitutes the transversal component of $b$.

The three unit vectors considered in that order, define a local right-handed orthogonal reference system associated with the pair of the polar vectors $r$ and $b$. The multiplication between the unit elements $e_r, e_t, e^s$ is governed by the multiplication table 1.1.3.

$$e_r = \frac{r}{|r|}, \quad e_t = \frac{b_t}{|b_t|}, \quad e^s = \frac{b^s}{|b^s|} \tag{1.1.40c}$$

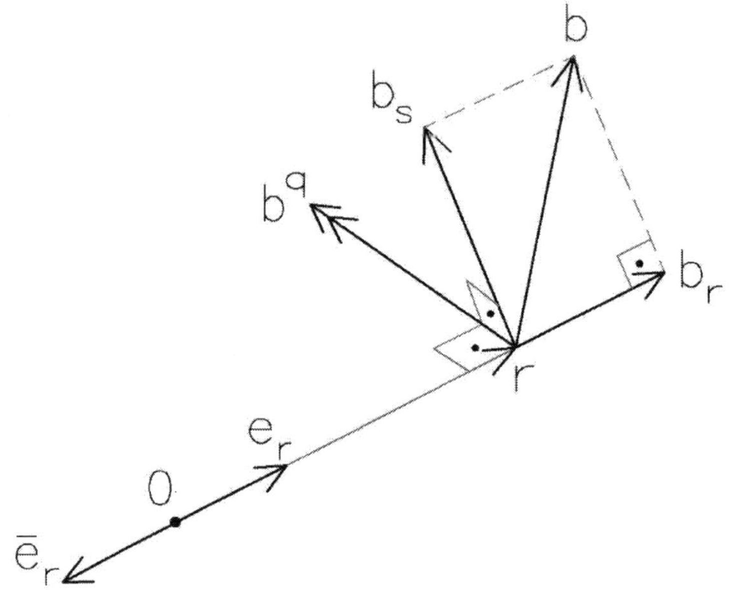


*Figure 1.1.4 - The free polar vector*

**Table 1 - Multiplication Table of $e_1$, $e_r$, $e_t$, $e^s$**

| | $e_1$ | $e_r$ | $e_t$ | $e^s$ |
|---|---|---|---|---|
| $e_1$ | 1 | $e_r$ | $e_t$ | $e^s$ |
| $e_r$ | $e_r$ | $\bar{1}$ | $e^s$ | $\bar{e}_t$ |
| $e_t$ | $e_t$ | $\bar{e}^s$ | $\bar{1}$ | $e_r$ |
| $e^s$ | $e^s$ | $e_t$ | $\bar{e}_r$ | $\bar{1}$ |
| | | | | |

### 1.1.9 The Polar Position

In the next chapters will become apparent that pairs consisting of a position vector equiangular with respect to the coordinate axes and a free polar vector $b$ applied at the position $r$ and intersecting the $z$-axis, furthermore referred as the *polar position* of the pair, have an important contribution to the investigation of the deformable continuous media. Any pair of vectors $r$ and $b$ with coordinates global reference system is transferred to the polar position after applying an orthogonal transformation furthermore referred as *polar transformation*, consisting of two successive rotations around axes passing through the origin of the global system (Figs. 1.1.5a, b)

By the first rotation the unit position vector $e_r$ is transferred to the unit equiangular vector

$$ {}_1e_r = \frac{1}{\sqrt{3}}\left(e_x + e_y + e_z\right) $$

and the free polar vector $b$ is transferred to the vector

$$ {}_1b = {}_1qb\ {}_1\tilde{q} \qquad (1.1.41) $$

where

$$ {}_1q = cos\frac{{}_1\hat{\phi}}{2} + {}_1u^a\ sin\frac{{}_1\hat{\phi}}{2} \qquad (1.1.41a) $$

the Hamiltonian quaternion defining the space rotation with

$$ {}_1u^a = (e_r \times_1 e_r)^a \qquad (1.1.41b) $$

$$ cos\ {}_1\hat{\phi} = -\left(e_r \cdot {}_1e_r\right)^{\bar{1}}, $$

$$ cos\frac{{}_1\hat{\phi}}{2} = \sqrt{\frac{1 - e_r \cdot {}_1e_r}{2}}, \qquad (1.1.41c) $$

$$sin\frac{{}_1\hat{\Phi}}{2} = \sqrt{\frac{1 + e_r \cdot {}_1e_r}{2}}$$

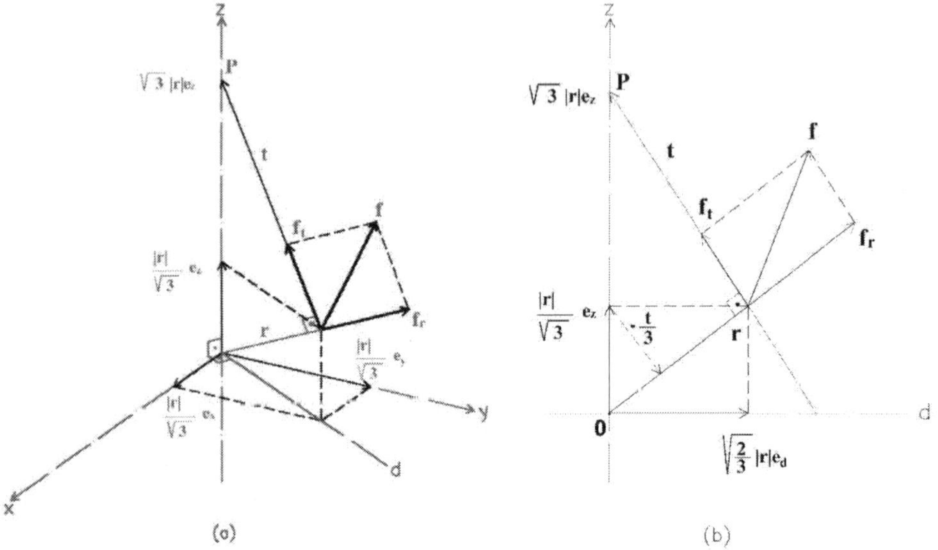


*Figure 1.1.5 - The Polar System*

The vector $e^{1s} = ({}_1r \times {}_1b)/|{}_1r \times {}_1b|$ is normal to the plane of ${}_1r$ and ${}_1b$, and the vector $e^s\ {}_1e_r \times e_z = \frac{1}{\sqrt{2}}(e_x - e_y)$ is normal to the plane of ${}_1e_r \times e_z$. By a second rotation around the ${}_1e^r$ axis at an angle ${}_2\hat{\Phi} = \arccos(e^{1s} \cdot e^{\bar{s}})$

the two vectors coincides and the transformed ${}_2b$ of ${}_1b$ intersects the $z$ −axis. It is

$${}_2b = {}_2q\ {}_1b\ {}_2\tilde{q} = {}_2q\ {}_1qb\ {}_1\tilde{q}\ {}_2\tilde{q} = qb\tilde{q} \qquad (1.1.42)$$

where

$${}_2q = cos\frac{{}_2\hat{\Phi}}{2} + {}_1e^r sin\frac{{}_2\hat{\Phi}}{2}, \qquad q = {}_2q\ {}_1q \qquad (1.1.42a)$$

In the products with the Hamiltonian quaternions the relations (1.2.34a) have to be taken into account.

The position of the pair of vectors $r$ and $b$ resulting from this transformation, which will be presented furthermore without the low left indices 1 or 2, will be referred as the *polar*. The intersection of the vector ${}_2b_t$ with the $z$ −axis at the point ${}_pz$ where, $|{}_pz| = 3|z| = \sqrt{3}|r|$, is the *pole* ${}_zP$ of the polar position. The vector

$$t = -\sqrt{2}e^s r = r - {}_pz = x + y - 2z = \frac{|r|}{\sqrt{3}}(e_x + e_y - 2e_z) \qquad (1.1.43)$$

connecting the pole ${}_zP$ with the position $r$ together with the vectors $r$ and $e^s$, are the characteristic vectors of the polar position. (Fig. 1.1.5)

In the polar position the axial vectors $r^a$ and ${}_ot^a$ of the area coordinates, corresponding to the polar vectors $r$ and $t$ respectively, are

$$r^a = (yz)^x + (zx)^y + (xy)^z = \frac{|r|^2}{3}(e^x + e^y + e^z) = \frac{|r|}{\sqrt{3}}(re)^a \quad (1.1.44a)$$

$${}_ot^a = (t_y t_z)^x + (t_z t_x)^y + (t_x t_y)^z = \frac{|r|^2}{3}(-2e^x - 2e^y + e^z) \quad (1.1.44b)$$

The vector $r^a$ is collinear with the vector $r$, but the vector ${}_ot^a$ is not collinear with the vector $t$ . Collinear with the vector $t$ is the axial vector

$$t^a = {}_ot^a + r^a = \frac{|r|^2}{3}(-e^x - e^y + 2e^z) = \frac{|r|}{\sqrt{3}}(et)^a \qquad (1.1.44c)$$

The Eqs. (1.1.44) and (1.1.45a, c) defining the vectors $t, r^a$ and $t^a$ in the polar position of the pair of vectors $r$ and $b$, are expressed in terms of the vectors $r$ and $e^s = \frac{(r\times b)}{|r\times b|}$ and can be applied to any other coordinate system.

In the polar position the totality of the vectors *r, b, t,* $r^a$ and $t^a$ are situated on the plane having $e^s$ as normal vector and contains the *z* and *d* axes, where $e_d = (e_x + e_y)/\sqrt{2}$ is the bisectrice of $e_x$ and $e_y$, which will be referred as the *polar plane* of the pair of vectors r and b. In (Fig. 1.1.5) the vector s and the poles ${}_zP$, ${}_dP$ on the polar plane are indicated together with their mirror images $r', t', r^{a\prime}, t^{a\prime}$ with respect to the z and d axes with the appropriate sign defining their orientations which are represented by the following, easily verified, relations

$$r' = \frac{|r|}{\sqrt{3}}(-e_x - e_y + e_z), \quad t' = -\frac{|r|}{\sqrt{3}}(e_x + e_y + 2e_z) \tag{1.1.45}$$

$$(-r)^a = r^a, \quad (-t)^a = t^a \tag{1.1.46a}$$

$$r'^a = (-r)^a = \frac{|r|^2}{3}(-e^x - e^y + e^z),$$
$$t'^a = (-t')^a = \frac{|r|^2}{3}(-e^x - e^y + e^z) \tag{1.1.46b}$$

The components of the vector $b$ in the reference systems *(s, r, -t)* and *(s, d, z)* are:

$$b_r = (b \cdot \hat{r})_1 r, \qquad b_t = (b \cdot \hat{t})_1 t \tag{1.1.47a}$$
$$b_d = (b \cdot \bar{e}_d)_1 e_d, \qquad b_z = (b \cdot \bar{e}_z)_1 e_z \tag{1.1.47b}$$

### 1.1.10 Orientation Vectors

An isolated system of points, as for example the material points constituting a solid, is represented geometrically through the reference of its points to a local orthogonal coordinate system having origin at a point, not necessarily belonging to the system, and an arbitrary basis consisting of the unit vectors $e_{x'}, e_{y'}, e_{z'}$. The position of the isolated system of points with respect to a global coordinate system, having as basis the unit vectors $e_x, e_y, e_z$ is determined by the position vector of the origin of the local coordinate system and the orientation of the local basis with respect to the global one. This orientation is determined by the orthogonal transformation from the global to the local reference system, which can be represented by one of the available sets of the three rotation parameters defining the transformation. As more appropriate is selected the first of the sets of rotation parameters, mentioned in section 1.1.6, consisting of a rotation angle $\hat{\phi}$ around an axis in space determined by the axial unit vector $e^{\phi}$ having equal direction cosines $e^{\phi i} = e^{\phi i'}$ in both systems, the local and the global.

The axial vector

$$\phi = \hat{\phi} e^{\phi} \qquad (1.1.48)$$

is the *orientation vector* of the local with respect to the global basis.

By the rotation $\phi$ the unit vectors $e_i$ of the global basis are transferred to the unit vectors $e_{i'}$ of the local basis having values, according to the Hamiltonian relation (1.1.35).

$$e_{i'} = q e_i \hat{q} \qquad (1.1.49)$$

where

$$q = cos\frac{\hat{\phi}}{2} + e^{\phi} sin\frac{\hat{\phi}}{2}, \quad \tilde{q} = cos\frac{\hat{\phi}}{2} - e^{\phi} sin\frac{\hat{\phi}}{2}, \quad q\tilde{q} = 1 \qquad (1.1.49a)$$

Inversely, for a given transformed basis consisting of the unit vectors $e_{i'}$, the orientation vector $\phi$ results according to the following procedure. Eq. (1.1.57) is written in the from

$$e_{i'} q - q e_i = 0 \quad \text{or} \quad (e_{i'} q) e^{\phi} - (q e_i) e^{\phi} = 0 \qquad (1.1.50)$$

After substituting Eq. (1.1.51), Eq. (1.1.50) becomes

$$(e_{i'} - e_i) e^{\phi} cos\frac{\phi}{2} - (e_{i'} + e_i) sin\frac{\phi}{2} =$$
$$= \left( (e_{i'} - e_i) \circ e^{\phi} cos\frac{\hat{\phi}}{2} \right)_{\bar{e}} + \left( (e_{i'} - e_i) \otimes e^{\phi} cos\frac{\hat{\phi}}{2} - (e_{i'} + e_i) sin\frac{\hat{\phi}}{2} \right)_p$$
$$= 0 \qquad (1.1.51a)$$

Eq. (1.1.51a) is satisfied for zero values of both, the imaginary and the vector part, resulting to the equations

$$(e_{i'} - e_i) \circ e^{\phi} = 0 \qquad (1.1.51b)$$

$$tan\frac{\dot{\phi}}{2} = \left((e_{i'} - e_i) \otimes e^{\phi}\right) (\widehat{e_{\iota'} + e_{\iota}}) \qquad (1.1.51c)$$

Eq. (1.1.52b) represents an homogeneous system of three linear algebraic equations with unknowns the three directions cosines $e^{\phi\bar{\iota}}$ of the unit vector $e^{\phi}$, having zero determinant and hence an infinity set of solutions.

With the additional condition

$$\sum_i \left(e^{\phi i}\right)^2 = 1 \qquad (1.1.51d)$$

result two solutions corresponding to two collinear unit vector $e^{\phi}$ with opposite directions. For the two opposite unit vectors result from Eq. (1.1.52c) two equal and opposite values respectively for the rotation angle $\hat{\phi}$, so that to both pairs corresponds the single orientation vector $\phi = \hat{\phi}\, e^{\phi}$

The sum ${}_{12}\phi = {}_{1}\phi + {}_{2}\phi$ of two orientation vectors ${}_{1}\phi$ and ${}_{2}\phi$ corresponds to the superposition of two rotations defined by the quaternions ${}_{1}q$ and ${}_{2}q$ satisfying the relation

$$p' = {}_{2}q\ {}_{1}q p\ {}_{1}\tilde{q}\ {}_{2}\tilde{q} = {}_{12}q p\, {}_{12}\tilde{q} \qquad (1.1.52)$$

According to the relation (1.1.49a) is

$$ {}_{12}q = {}_{2}q\ {}_{1}q = cos\frac{2\hat{\phi}}{2}cos\frac{1\hat{\phi}}{2} + e^{\phi}cos\frac{1\hat{\phi}}{2}sin\frac{2\hat{\phi}}{2} + e^{1\phi}sin\frac{1\hat{\phi}}{2}cos\frac{2\hat{\phi}}{2} + e^{2\phi}e^{1\phi}sin\frac{1\hat{\phi}}{2}sin\frac{2\phi}{2} \qquad (1.1.53)$$

In the right side of Eq. (1.1.54) is contained the non-commutative term $e^{2\phi}e^{1\phi}$ . This implies that the superposition of rotations is not commutative and the field of the orientation vectors is not commutative in addition, so that

$$ {}_{12}\phi = {}_{12}\hat{\phi}\, e^{12\phi} = {}_{1}\phi + {}_{2}\phi \neq {}_{2}\phi + {}_{1}\phi = {}_{21}\phi = {}_{21}\hat{\phi}\, e^{21\phi} \qquad (1.1.54)$$

Eq. (1.1.54), after the following symbolic extension of the quaternion product expression, can be put under the form

$$ {}_{12}q = {}_2q\ {}_1q = {}_2q \cdot {}_1q + {}_2q \times {}_1q \tag{1.1.55}$$

where

$$ {}_2q \cdot {}_1q = \cos\frac{{}_{12}\widehat{\Phi}}{2} = \cos\frac{{}_2\widehat{\Phi}}{2}\cos\frac{{}_1\widehat{\Phi}}{2} + \left(e^{2\Phi}\sin\frac{{}_2\widehat{\Phi}}{2}\right)\cdot\left(e^{1\Phi}\sin\frac{{}_1\widehat{\Phi}}{2}\right) \tag{1.1.55a}$$

$$ q_2 \times q_1 = e^{12\Phi}\sin\frac{{}_{12}\widehat{\Phi}}{2} = e^{2\Phi}\sin\frac{{}_2\widehat{\Phi}}{2}\cos\frac{{}_1\widehat{\Phi}}{2} + e^{1\Phi}\sin\frac{{}_1\widehat{\Phi}}{2}\cos\frac{{}_2\widehat{\Phi}}{2} + \left(e^{2\Phi}\sin\frac{{}_2\widehat{\Phi}}{2}\right)\times\left(e^{1\Phi}\sin\frac{{}_1\widehat{\Phi}}{2}\right) \tag{1.1.55b}$$

$$ {}_{12}\widehat{\Phi} = 2arc\,\dot{\cos}(\,{}_2q \cdot {}_1q), \qquad e^{12\Phi} = \frac{{}_2q \times {}_1q}{\sin\frac{{}_{12}\widehat{\Phi}}{2}} \tag{1.1.55c}$$

In the case of infinitesimal, instead of finite, rotations $\Delta\phi$ it is $\cos\frac{\Delta\widehat{\phi}}{2} \cong 1$, $\sin\frac{\Delta\widehat{\phi}}{2} \cong \frac{\Delta\widehat{\phi}}{2}$ and the non-commutative term in Eq. (1.1.52), which is multiplied by second order infinitesimals, vanishes. As a result the superposition of infinitesimal rotations is commutative and the corresponding orientation vector satisfies the relation:

$$ \Delta_{12}\phi = \Delta\,{}_2\phi + \Delta\,{}_1\phi = \Delta\,{}_1\phi + \Delta\,{}_2\phi \tag{1.1.56}$$

The angular velocities representing infinitesimal rotations executed at infinitesimal time intervals are likewise commutative in addition.

## 1.2 OCTONION ANALYSIS

### 1.2.1 Octonion Functions

An octonion

$$f = \sum_I f_I = \sum_I \hat{f}_I e_I$$

is called an *octonion function* of the octonion arguments or variables

$${}_m u = \underset{J}{\subset_m} u_J = \underset{J}{\subset_m} \hat{u}_J e_J, \qquad {}_n u = \underset{K}{\subset_n} u_K = \underset{K}{\subset_n} \hat{u}_K e_K, \dots\dots,$$

if to every set of numerical values that the scalar coefficients ${}_m\hat{u}_J, \dots\dots\, {}_n\hat{u}_K$ of the components ${}_m u_{J,\dots\dots} {}_n u_{K,\dots\dots}$ of the arguments ${}_m u_{,\dots\dots} {}_n u_{,\dots\dots}$ can takc on, there correspond a definite set of values of the scalar coefficients $\hat{f}_I$ of the components $f_I$of $f$. In the present notation the arguments of an octonion function will be included in brackets [ ] and not in parentheses, which are destined exclusively for the notation of two-term products binomials, according to the expression

$$f\left[\underset{m}{\subset_m} u\right] = \sum_I f_I\left[\underset{J}{\subset_m} u_J, \underset{K}{\subset_n} u_K, \dots\dots\right] = \\ = \sum_I \left(\hat{f}_I\left[\underset{J}{\subset_m} \hat{u}_J e_J, \underset{K}{\subset_n} \hat{u}_K e_K, \dots\dots\right] e_I\right) \qquad (1.2.1)$$

In the sequence of this presentation special consideration will be reserved for octonion functions

$$f[u_1, r] = f[u_1, x, y, z] = \sum_I f_I[u_1, x, y, z]$$

having arguments a scalar component $u_1$ and $a$ position vector $r = x + y + z$, which will be used for the mathematical manipulation of the main subject of this treatise.

### 1.2.2 Partial derivatives

An octonion function

$$f = \sum_I f_I$$

has a partial derivative (or is differentiable) with respect to one of its argument components $u_J$, which is denoted by the symbol $f_{,\hat{J}}$ if there exists the limit of the quotient of the differentials $_J(\partial f)$ and $\partial u_J$ expressed in the binomial form two-term product form

$$f_{,\hat{J}} = {}_J(\partial f)(\widehat{\partial u_J}) = \lim_{\Delta u_J \to 0}(f[u_J + \Delta u_J] - f[u_J])\,(\widehat{\Delta u_J}) \qquad (1.2.2)$$

and has the same value for any mode of the approach $\Delta u_J \to 0$. The second factor of the Eq. (1.3.2) has the form

$$\widehat{\Delta u_J} = -\frac{\Delta u_J}{|\Delta u|^2}$$

which implies that the derivative is not single-valued, because due to the division by $|\Delta u|^2$, it depends on the variations of the totality of the components of the argument $u$ and not exclusively on $\Delta u_J$. The partial derivative $f_{,\hat{J}}$ is single valued in the case $|\Delta u|^2 = |\Delta u_J|^2$, $\Delta u_K = 0\ for\ K \neq J$, and the second factor of the derivative binomial consists of the reverse argument component, so that

$$f_{,\hat{J}} = f_{,\bar{J}} = {}_J(\partial f)(\overline{\partial u_J})$$

This definition permits the differential $_J(\partial f)$ to be considered equivalent to the expression

$$_J(\partial f) = f[u_J + \partial u_I] - f[u_J] \qquad (1.2.3)$$

for infinitesimal $\partial u_J$ and considered only as the first factor of a partial derivative product. From the expressions (1.3.2) and (1.3.3) results

$$f[u_J + \partial u_J] = f[u_J] + f_{,\bar{J}}(\partial u_J) \qquad (1.2.4)$$

It is easily verified that

$$f_{,\bar{J}} = \left(\sum_I f_I\right)_{,\bar{J}} = \sum_I f_{I,\bar{J}} = \sum_I (\partial f_I)(\overline{\partial u_J}) \tag{1.2.5}$$

Where $f_{I,\bar{J}} = (\partial f_I)(\overline{\partial u_J})$ are the partial derivatives with respect to $u_J$ of the components $f_I$ of $f$. The symbols $f_{I,\bar{J}}$, representing a two term product of octonion components, is purposeful to follow the general notation introduced for two term product and to be included in a parenthesis under the form $\left(f_{I,\bar{J}}\right)_K$ where $K$ is the index of the $e_K$ corresponding to the unit vectors $e_I$ and $\bar{e}_J$ according to the multiplication table 1.2.2. For example the partial derivative of the octonion function component $f_x$ with respect to $y$ have to be written $\left(f_{x,y}\right)^{\bar{z}}$. The partial derivative of $f_I$ with respect to a real scalar argument $u_1$ has the form $\left(f_{I,1}\right)_I$. The partial derivative with respect to the argument component $u_K$ of the product of two octonion function components $f_I$ and $f_J$, using the relation (3.4) and considering the second order infinitesimal $(\partial u_K)^2$ as negligible, results as follows :

$$\begin{aligned}
\left(f_I f_J\right)_{,K} &= \left(\partial\left(f_I f_J\right)\right)\left(\overline{\partial u_K}\right) = \left(\left(f_I f_J\right)[u_K + \partial u_K] - \left(f_I f_J\right)[u_K]\right)\left(\overline{\partial u_K}\right) = \\
&= \left(f_I[u_K + \partial u_K] f_J[u_K + \partial u_K] - f_I[u_K] f_J[u_K]\right)\left(\overline{\partial u_K}\right) = \\
&= \left(\begin{array}{l}\left(f_I[u_K] + f_{I,\bar{K}}(\partial u_K)\right)\left(f_J[u_K] + f_{I,\bar{k}}(\partial u_K)\right) \\ -f_I[u_K] f_J[u_K]\end{array}\right)\left(\overline{\partial u_K}\right) = \\
&= \left(\left(f_{I,\bar{K}}(\partial u_K)\right) f_J + f_I\left(f_{J,\bar{K}}(\partial u_K)\right)\right)(\partial u_K)
\end{aligned}$$

or

$$\left(f_I f_J\right)_{,\bar{K}} = \theta_{JK} f_{I,\bar{K}} f_J + \theta_{IK} f_I f_{J,\bar{K}} \tag{1.2.6}$$

A second order partial derivative of an octonion function component $f_I$ with respect to two argument components $u_J$ and $u_K$, executed in that order, is denoted by the symbol $f_{I,\overline{JK}}$, which is an abridged expression of the symbol $\left(f_{I,\bar{J}}\right)_{,\bar{K}}$. It is evident that

$$f_{I,\overline{JK}} = \theta_{JK} f_{I,\overline{KJ}} \tag{1.2.7}$$

Generally, higher order partial derivatives of $f_I$ with respect to argument components $u_J, u_K, u_i$ ..... are denoted by the symbol $f_{I,\overline{JKL}}$ which is equivalent to the expression $\left(\left(f_{I,\bar{J}}\right)_{,\bar{K}}\right)_{,L}$, ....

### 1.2.3 Derivatives with respect to an octonion

An octonion function

$$f[u] = \sum_I f_I[u]$$

of an octonion variable

$$u = \sum_I u_I$$

has a derivative (or it is differentiable) with respect to $u$, which is denoted by the symbol $f,\hat{u}$, ıf there exists the limit of the quotient of the differentials $\partial f$ and $\partial u$ expressed in the product form

$$f_{,\hat{u}} = (\partial f)(\widehat{\partial u}) = \lim_{\Delta u \to 0}\left((f[u+\Delta u] - f[u])\widehat{\Delta u}\right) =$$

$$= \sum_I \underset{J}{\subset} \lim_{\Delta u \to 0}\left(\left(f_I\left[\underset{J}{\subset}(u+\Delta u_J)\right] - f_I[u]\right) = \Delta u_J\right) \qquad (1.2.8)$$

and has the same value for any mode of approach to $u$, from any point $u + \Delta u$ lying in its neighborhood. If $f[u]$ has a derivative for an octonion $u$ and also for any other octonion in the neighborhood of $u$ is said to be an *analytic octonion function* at $u$.
According to Eq. (1.2.6) the derivative with respect to an octonion argument $u$ of the product of two analytic functions $f$ and $g$ will be

$$(fg)_{,\bar{u}} = \left(\sum_I \sum_J f_I g_J\right)_{,\bar{u}}$$

$$= \sum_I \sum_J \sum_K (f_I g_J)_{,\bar{K}} = \sum_I \sum_J \sum_K \left(\theta_{JK} f_{I,\bar{K}} \theta_{IK} + f_I g_{J,\bar{K}}\right) \qquad (1.2.9)$$

The derivative of a composite analytic function $f[g]$, where $g = g[u]$, is likewise analytic with respect to the octonion argument $u$, will be

$$
\begin{aligned}
f[g[u]]_{,\bar{u}} &= (\partial f)(\overline{\partial u}) = \left((\partial f)\left((\widehat{\partial g})(\partial g)\right)_1\right)(\overline{\partial u}) = \\
&= \left(\left((\partial f)(\overline{\partial g})\right)(\partial g)\right)(\overline{\partial u}) = \left(f_{,\bar{g}}(\partial g)\right)(\overline{\partial u}) = \\
&= \sum_I \sum_J \sum_K \theta_{IJK}\left(f_{,\bar{g}}\right)_I \left((\partial g_J)(\overline{\partial u_K})\right) = \\
&= \sum_I \sum_J \sum_K \theta_{IJK}\left(f_{,\bar{g}}\right)_I (g_{J,\bar{K}}
\end{aligned}
\qquad (1.2.10))
$$

### 1.2.4 Extended Cauchy – Riemann Equations

Necessary conditions for $f[u]$ to have a derivative $f_{,\hat{u}}$ for any $\Delta u \to 0$, are the derivative to exist and to have the same value for any one of the eight modes $\Delta u = \Delta u_J$ and $\Delta u_K = 0$ for $K \neq J$. For each one of these eight modes the derivative will be

$$
\begin{aligned}
f_{,\hat{u}} = f_{,\bar{J}} &= \sum_I \left(f_{I,\bar{J}}\right)_K = \sum_I (\partial \hat{f}_I e_I)(\partial \hat{u}_J e_J) = \\
&= \sum_I \left(f_{I,J}\right)_K \left(e_I \bar{e}_J\right)_K = \sum_I \left(f_{I,J}\right)_K e_K
\end{aligned}
\qquad (1.2.11)
$$

and will have the same value if the $K$ –components of the eight expressions of $f_{,\hat{u}}$ are equal. According to the multiplication table, to each one index $K$ correspond eight pairs of indices $I, J$, implying the following equations :

For $e_K = e_1$ the 7 equations

$$\hat{f}_{1,\bar{1}} = \hat{f}^{e,\bar{e}} = \hat{f}_{i,\bar{\imath}} = \hat{f}^{i,\bar{\imath}} \qquad (1.2.12a)$$

For $e_K = e$, the 7 equations

$$\hat{f}^{e}{}_{,\bar{1}} = -\hat{f}_{1,}^{\bar{e}} = \hat{f}_{i,}^{\bar{\imath}} = -\hat{f}^{i}{}_{,\bar{\imath}} \qquad (1.2.12b)$$

For $e_K = e_k$ the 21 equations

$$\hat{f}_{k,\bar{1}} = -\hat{f}_{1,\bar{k}} = \hat{f}^{k,\bar{e}} = -\hat{f}^{e,k} = \hat{f}^{j}{}_{,\bar{\imath}} = -\hat{f}_{i,}{}^{\bar{J}} = \hat{f}_{j,}{}^{\bar{\imath}} = -\hat{f}^{i}{}_{,\bar{J}} \qquad (1.2.12c)$$

For $e_K = e^k$ the 21 equations

$$\hat{f}^{k,}{}_{\bar{1}} = -\hat{f}_{1}{}^{,\bar{k}} = \hat{f}^{e}{}_{,\bar{k}} = -\hat{f}_{k,}{}^{\bar{e}} = -\hat{f}_{j,\bar{\imath}} = \hat{f}_{i,\bar{J}} = -\hat{f}^{i,\bar{J}} = -\hat{f}^{j,\bar{\imath}} \qquad (1.2.12d)$$

where $i, j, k,$ are cyclic permutations of $x, y, z$.

The 56 differential equations (1.2.10) relating the real functions constituting the scalar coefficients of the components of the octonion function $f$ constitute the octonion extension of the Cauchy-Riemann equations established for the real coefficients of the components of the complex analytic functions.

The above conditions (1.2.10) are also sufficient, at least if the partial derivatives are continuous functions of $u_{\wp}$ as may easily be shown by writing

$$\hat{f}_{\bar{I}}\left[\subset_J(\hat{u}_J + \Delta\hat{u}_J)\right] = \hat{f}_{\bar{I}}\left[\subset_J \hat{u}_J\right] + \sum_J \hat{f}_{\bar{I},\bar{J}}\Delta\hat{u}_J + \text{higher order infinitesimals}$$

inserting them in Eq. (1.2.8) and then using Eq. (1.2.10).
Under the above conditions the second factor of the derivative binomial is the reverse of the differential $\partial u$ and the derivative is

$$f_{,\hat{u}} = (\partial f)(\overline{\partial u}) = f_{,\bar{u}} \qquad (1.2.13)$$

### 1.2.5 Octonion Potential Functions

Considering for the analytic octonion functions $f[u]$ and the octonion argument $u$ the expressions

$$f[u] = f_1[u] + f_p[u] + f^a[u] + f^e[u], \qquad u = u_1 + u_p + u^a + u^e$$

the derivative $f_{,\bar{u}}$ takes the form

$$f_{,\bar{u}} = (f_{,u})_1 + (f_{,\bar{u}})_p + (f_{,\bar{u}})^a + (f_{,\bar{u}})^e \tag{1.2.14}$$

where

$$(f_{,\bar{u}})_1 = f_{1,\bar{1}} + f_{p,\bar{p}} + f^{a,\bar{a}} + f^{e,\bar{e}}$$

$$(f_{,\bar{u}})_p = f_{1,\bar{p}} + f_{p,\bar{1}} + f_{p,}{}^{\otimes\bar{a}} + f^{a}{}_{,\otimes\bar{p}} + f^{a,\bar{e}} + f^{e,\bar{a}}$$

$$(f_{,\bar{u}})^a = f_{1,}^{\ \bar{a}} + f^{a}{}_{,\bar{1}} + f_{p,\times\bar{p}} + f^{a,\times\bar{a}} + f_{p,}{}^{\bar{e}} + f_{,\bar{p}}{}^{e}$$
$$(f_{,\bar{u}})^e = f_{1,}{}^{\bar{e}} + f_{p,}{}^{\circ\bar{a}} + f^{a}{}_{,\circ\bar{p}} + f^{e,\bar{1}}$$

and, according to Eq. (1.2.7),

$$f_{p,\cdot\bar{p}} = (\partial f_p)\cdot(\partial\bar{u}_p) = \sum_i f_{i,\bar{\imath}} \qquad f^{a,\cdot\bar{a}} = (\partial f^a)\cdot(\overline{\partial u^a}) = \sum_i f^{i,\bar{\imath}}$$

$$f_p{}^{\circ\bar{a}} = (\partial f_p)\circ(\overline{\partial u^a}) = \sum_i f_{i,}{}^{\bar{\imath}} \qquad f^{a}{}_{,\circ\bar{p}} = (\partial f^a)\circ(\overline{\partial u_p}) = \sum_i f^{i}{}_{,\bar{\imath}}$$

$$f_{p,\times\bar{p}} = (\partial f_p)\times(\partial\bar{u}_p) = \sum_i\sum_{j\neq i} f_{i,\bar{\jmath}} \qquad f^{a,\times\bar{a}}(\partial f^a)\times(\overline{\partial u^a}) = \sum_i\sum_{j\neq i} f^{i,\bar{\jmath}}$$
$$f_p{}^{\otimes\bar{a}} = (\partial f_p)\otimes(\overline{\partial u^a}) = \sum_i\sum_{j\neq i} f_{i,}{}^{\bar{\jmath}} \qquad f^{a}{}_{,\otimes\bar{p}}(\partial f^a)\times(\overline{\partial u_p}) = \sum_i\sum_{j\neq i} f^{i}{}_{,\bar{\jmath}}$$

In the case the argument $u$ is the position vector $r$, so that $u = u_p = r,\ u_1 = u^a = u^e$ and taking into consideration the symbols introduced in the ordinary vector analysis, the components of the derivative $f_{,\bar{r}}$ will be

$$(f_{,\bar{r}})_1 = f_{p,\cdot\bar{r}} = f_p\cdot\nabla = div\, f_p$$
$$(f_{,\bar{r}})_p = f_{1,\bar{r}} + f^{a}{}_{,\otimes\bar{r}} = f_1\nabla + f^a\otimes\nabla = grad\, f_1 + grad\, f^a$$
$$(f_{,\bar{r}})^a = f_{p,\times\bar{r}} + f^{e}{}_{,\bar{r}} = f_p\times\nabla + f^e\nabla = curl\, f_p + curl\, f_e$$

$$\left(f_{,\bar{r}}\right)^{e} = f^{a}{}_{,\circ\bar{r}} = f^{a} \circ \nabla = div\, f^{a}$$

where

$$\nabla = \partial\left(\overline{\partial r}\right) = \partial\left(\overline{\partial x}\right) + \partial\left(\overline{\partial y}\right) + \partial\left(\overline{\partial z}\right) \qquad (1.2.15a)$$

the *nabla* differential operator, with the symbol $\partial$ having the character of a real scalar, so that

$$f;\nabla = f;\left(\partial\left(\overline{\partial r}\right)\right) = (\partial f);\left(\overline{\partial r}\right) = f_{;\bar{r}} \qquad (1.2.15b)$$

In (1.2.15b) the symbol; represents any type of product (dot, cross, hole or grid).

A generalization of the above notation for the case of an octonion argument $u$ can be obtained by the introduction of the differential operator with components

$$\rhd = \partial(\partial u) = \partial\left(\overline{\partial u_1}\right) + \partial\left(\overline{\partial u_p}\right) + \partial\left(\overline{\partial u^a}\right) + \partial\left(\overline{\partial u^e}\right) \qquad (1.2.16)$$

Eq. (1.2.14) can be represented in the simplified form

$$\left(f_{,\bar{u}}\right)_1 = f \cdot\rhd = \begin{Bmatrix} \partial f_1 & \partial f_p & \partial f^a & \partial f^e \\ \downarrow & \downarrow & \downarrow & \downarrow \\ \overline{\partial u_1} & \overline{\partial u_p} & \overline{\partial u^a} & \overline{\partial u^e} \end{Bmatrix} \qquad (1.2.17a)$$

$$\left(f_{,\bar{u}}\right)_p = f \otimes \rhd = \begin{Bmatrix} \partial f_1 & \partial f_p & \partial f^a & \partial f^e \\ \text{⤩} & \text{⤩} & \text{⤩} & \\ \overline{\partial u^e} & \overline{\partial u^a} & \overline{\partial u_p} & \overline{\partial u_1} \end{Bmatrix} \qquad (1.2.17b)$$

$$\left(f_{,\bar{u}}\right)^a = f \times\lhd = \begin{Bmatrix} \partial f_1 & \text{⤩} & \partial f_p & \text{⤩} & \partial f^a & \text{⤩} & \partial f^e \\ \overline{\partial u^e} & & \overline{\partial u^a} & & \overline{\partial u_p} & & \overline{\partial u_1} \end{Bmatrix} \qquad (1.2.17c)$$

$$\left(f_{,\bar{u}}\right)^e = f \circ\lhd = \begin{Bmatrix} \partial f_1 & \partial f_p & \partial f^a & \partial f^e \\ \downarrow & \downarrow & \downarrow & \downarrow \\ \overline{\partial u^e} & \overline{\partial u^a} & \overline{\partial u_p} & \overline{\partial u_1} \end{Bmatrix} \qquad (1.2.17d)$$

The symbol $\lhd$ contains the terms $\partial u_1$ of the symbol $\rhd$ taken in the inverted order. The expression in { } represent the sums of the two - term products indicated by the arrows, which are of the type defined by the symbol following $/$ in the preceding abridged expression. More briefly, for parallel arrows the products are of the dot or hole type, while for crossing arrows the products are of the cross or grid type.

Finally the derivative $f_{,u}$ can be represented by the following expression

$$f_{,\bar{u}} = f \lhd\rhd \tag{1.2.18}$$

after the introduction of the differential operator

$$\lhd\rhd = (\cdot + \otimes) \rhd + (\times + \circ) \lhd \tag{1.2.19}$$

Under this form an analytic octonion function $f$ can be considered as an octonionic potential of an octonion field, with its derivatives $(f \cdot \rhd)_1$, $(f \otimes \rhd)_p$, $(f \times \lhd)^a$, $(f \circ \lhd)^e$ constituting the octonionic form of the real divergence, the gradient, the curl and the imaginary divergence respectively.

### 1.2.6 Octonion Line Integrals

The integral of a real function of a real variable is defined as the limit of a certain sum for all values of the variable taken over an interval of $R$. The same formal definition is carried over to the case of an octonion function of an octonion variable for values taken over an interval along a path of the variable. A path $C$ of the variable $\omega$ is represented by the parametric equation

$$\omega = \omega[u_1] = \sum_J \omega_J[u_1] = \sum_J \omega_J[u_1] e_J \tag{1.2.20}$$

where $u_1$ is a scalar parameter taking values in the interval between the end values ${}_1u_1$ and ${}_2u_1$ to which correspond the end points ${}_1\omega$ and ${}_2\omega$ of the path.

From (1.2.21) results

$$d\omega = \sum_J \left(\omega_{J,\bar{u}_1} du_1\right)_J = \sum_J \left(\omega_{j,\bar{u}_1} du_1\right)_1 e_J \qquad (1.2.21)$$

The differential $d\omega$ is not collinear to $\omega$, except the case all $\omega_J$are linear functions of $u_1$ without constant terms.

The integral of an octonion function $f[\omega]$ analytic or not, over the path $C$ will be

$$\int_C f[\omega]\, d\omega = \int_C \left(\sum_I f_I\right)\left(\sum_J d\omega_J\right) =$$

$$= \sum_I \sum_J \int_C f_I d\omega_J = \sum_I \sum_J \left(\int_C \hat{f}_I d\hat{\omega}_J\right)(e_I e_J) =$$

$$= \sum_I \sum_J \left(\int_{1u_1}^{2u_1} f_I(\hat{\omega}_{J,u_1})\, du_1\right)(e_I e_J) \qquad (1.2.22)$$

From (1.2.22) results that the integral is an octonion having ordinary integrals of real functions as scalar coefficients of the $(e_I e_J)$ - components.

### 1.2.7 Extended Cauchy's Integral Theorem

For a complex analytic function

$$f[C] = \hat{f}_1[C] + \hat{f}^e[C]e$$

of the complex variable

$C = \hat{C}_1 + \hat{C}^e e$ and $dC = d\hat{C}_1 + d\hat{C}^e e$

defined in a simply connected domain of $C$, the Cauchy's integral theorem states that its line integral along a closed path $S$ lying in the same domain satisfies the equation

$$\oint_S f[C]dC = \oint_S (\hat{f}_1[C] + \hat{f}^e[C]e)(d\hat{C}_1 + d\hat{C}^e) = 0 \qquad (1.2.23)$$

An octonion analytic function $f[\omega]$ of the octonion variable

$$\omega = \sum_i (\omega_i + \omega^i) = \sum_i (\widehat{\omega}_i e_i + \widehat{\omega}^i e^i) = \sum_i e_i(\widehat{\omega}_i + \widehat{\omega}^i e) = \sum_i e_i C_i$$
(1.2.24a)

as $e^i = e_i e$ and

$$d\omega = \sum_i e_i(d\widehat{\omega}_i + d\widehat{\omega}^i e) = \sum_i e_i dC_i \qquad (1.2.24b)$$

where $i = 1, x, y, z$ and for $i = 1, \widehat{\omega}^i = \widehat{\omega}^e$, which is defined likewise in a simply connected domain of $\omega$, has the form

$$f[\omega] = \sum_j (f_j[\omega] + f^i[\omega]) =$$
$$= \sum_j e_j(\hat{f}_j[\omega] + \hat{f}^j[\omega]e) = \sum_j e_j f_{Cj}[\omega] \qquad (1.2.25)$$

Eqs. (1.2.24, 25) define *primitive representation* of octonions in the basis of primitive unit elements $1, e_x, e_y, e_x$ with complex instead of real coefficients, the $C_i$ for the variable and $f_{Ci}[\omega]$ for the function.

The line integral of $f[\omega]$ a along a closed path $S$ contained in the same domain will satisfy the following equation which constitutes the Cauchy's integral theorem extended for octonion analytic functions.

$$\oint_S f[\omega]d\omega = \oint_S \left(\sum_j e_j f_{Cj}[\omega]\right)\left(\sum_i e_i dC_i\right) =$$

$$= \oint_S \left( \sum_j e_j (\hat{f}_j[\omega] + \hat{f}^j[\omega]) \right) \left( \sum_i e_i (d\widehat{\omega}_i + d\widehat{\omega}^i e) \right) =$$

$$= \sum_j \sum_i e_j e_i \oint_{{}_iS} (\hat{f}_j[\omega] - \hat{f}^j[\omega] e)(d\widehat{\omega}_i + \theta_{ji} d\widehat{\omega}^i e) = 0$$

(1.2.26)

since, according to the Eq. (1.2.23) each one integral of the sum equals zero. In Eq. (1.2.26) ${}_iS$ is the projection of the path $S$ on the plane of the unit elements $e_i$ and $e^i$. At each one integrand the components $\omega_i$ and $\omega^i$ of $\omega$ are considered as its arguments, which the rest six components are considered as parameters of the functions $\hat{f}_j$ and $\hat{f}^j$ with values the components of $\omega$ corresponding to the points of the path $S$ which are projected on the path ${}_iS$.

If the path $S$ is subdivided into two arcs ${}_1S$ and ${}_2S$, then Eq. (1.2.26) takes the form

$$\oint_S f[\omega]\, d\omega = \int_{{}_1S} f[\omega] d\omega + \int_{{}_2S} f[\omega] d\omega = 0$$

and, consequently

$$\oint_{{}_1S} f[\omega]\, d\omega = -\int_{{}_2S} f[\omega] d\omega = \int_{{}_2S^*} f[\omega] d\omega \qquad (1.2.27)$$

where ${}_2S^*$ is the arc ${}_2S$ traced in the opposite direction than that of the closed path $S$. Eq (1.2.27) holds for any paths joining any points ${}_1\omega$ and ${}_2\omega$ and lying entirely in the domain $D$ where $f[\omega]$ is analytic. Then the integral of $f[\omega]$ from ${}_1\omega$ to ${}_2\omega$ is independend of the path in $D$.

### 1.2.8 Extended Cauchy's Integral Formula

The Cauchy's integral formula for complex functions is

$$\oint_S f[c]\left(\widehat{c - {}_0c}\right)dc = \oint_S f[c]d\left(ln(c - {}_0c)\right) = 2\pi f\left[{}_0c\right]e \qquad (1.2.28)$$

where $f[c]$ is a function of the complex variable $c$ analytic in a simply connected domain $D$ of $c$, $\widehat{c - {}_0c}$ the inverse of $c - {}_0c$ and $S$ a closed path lying in the same domain $D$. The integrand is likewise analytic for all values of $c$ except the value ${}_0c$. Through the multiplication of $f\left[{}_0c\right]$ by $e$ in the right side of Eq. (1.2.28) a parity transformation is performed, since its real component is transformed to imaginary and the imaginary to real.

It is easily verified that, for any primitive unit vector $e_i$, it is

$$\left(\widehat{c - {}_0c}\right)dc = \left(\left(\widehat{c - {}_0c}\right)\bar{e}_i\right)(e_i dc) \qquad (1.2.29)$$

Eq. (1.2.28) can be extended for octonion functions $f[\omega]$ of the octonion variable $\omega$, which are analytic in a simply connected domain $D$ of $\omega$, in the following manner.

For an octonion $\omega$ represented by its primitive components according to Eq. (1.2.24a) its *primitive reverse* is defined by the expression

$$\breve{\omega} = \sum_i (\widehat{\omega_\iota + \omega^\iota}) = \sum_i (\widehat{e_\iota c_\iota}) = \sum_i \hat{c}_i \hat{e}_i, \qquad (i = 1, x, y, z) \qquad (1.2.30)$$

with components the inverses of the primitive components of $\omega$.

The following integral and its equivalent expression according to Eqs. (1.2.24a, b) and (1.2.30)

$$\oint_S f[\omega]\left(\left(\widehat{\omega - {}_0\omega}\right)d\omega\right) = \sum_k \oint_{{}_kS} \left(\sum_i e_i f_{ci}[\omega]\right)\left(\left(\sum_j \left(\widehat{c_j - {}_0c_j}\right)e_j\right)(e_k dc_k)\right)$$

is considered, where $f[\omega]$ is analytic in a simply connected domain $D$ of $\omega$, ${}_kS$ is the projection of the closed path $S$ on the plane of the unit vectors

$e_k$ and $e^k$, and ${}_0\omega$ an octonion in D with each one of its projections ${}_0\omega_k + {}_0\omega^k$ enclosed by ${}_kS$.

For each one term of the triple sum of integrals the components $e_k c_k$ of $\omega$ is considered as the variable of the functions $f_{ci}[\omega]$ while the rest three components are considered as parameters. For $j \neq k$ the corresponding integrands in the triple sum are analytic functions for all values of the variable $e_k c_k$ and according to the Cauchy's integral theorem, their integrals are equal to zero. The terms of the triple sum with non zero integral, are those corresponding to $j = k$, which are analytic for all values of $e_k c_k$ except the value $e_j\, {}_0c_j = e_k\, {}_0c_k$. The integral, after taking into consideration the Eqs. (1.2.28, 29), takes the form

$$\oint_S f[\omega]\left((\omega - {}_0\omega) d\omega\right)$$

or

$$\oint_S f[\omega]\left(\omega \widehat{-\ {}_0}\omega\right) d\omega = 2\pi f[\,{}_0\omega]e =$$

$$= \sum_j \oint_{{}_jS} \left(\sum_i e_i f_{ci}\right)\left(\left((c_j - {}_0c_j)\bar{e}_j\right)(e_j dc_j)\right) =$$

$$= 2\pi\left(\sum_i e_i f_{ci}\left[\sum_j e_j\, {}_0c_j\right]\right)e \qquad (1.2.31)$$

Eq. (1.2.31) is the Cauchy's integral formula extended for octonion analytic functions of an octonion variable. It is furthermore noticed that through the multiplication of $f[\,{}_0\omega]$ by $e$ in the right side of Eq. (1.2.31b) a parity transformation is performed since its primitive components are transformed to derived and vice – versa.

### 1.2.9 Formulas of Gauss, Stokes and Green

In this section are presented in the octonion notation the formulas of Gauss, Stokes and Green of the ordinary vector analysis, relating line, surface and volume integrals of scalar and polar vector functions $b_1[r]$ and $b_p[r]$

respectively, with argument the position vector $r = x + y + z$, and of their products with the differential operator $\nabla$ as defined in Eqs. (1.2.5).

If $A$ is a closed surface including the volume segment $V^e$, the *Gauss' formula* takes the form

$$\int_{V^e} \left((div\, b_p)_1 dV^e\right)_e = \int_{V^e} \left((b_p \cdot \nabla)_1 dV^e\right)^e = \int_{V^e} \left((b_{p \cdot \bar{r}})_1 dV^e\right)^e =$$

$$= -\int_A \left(b_p \circ dA^n\right)^{\bar{e}} \qquad (1.2.32)$$

For $r = x + y$ and $S$ a closed plane curve including the surface segment $A^z$ lying on the $(x, y)^z$- plane, the *First Green' s Formula*, which is equivalent to the Gauss Formula specified in the plane, takes the form

$$\int_{A^z} \left((div\, b_p)_1 dA^z\right)^{\bar{z}} = \int_{A^z} \left((b_p \cdot {}^z\nabla)_1 dA^z\right)^z = \int_{A^z} \left((b_{p,\cdot\bar{r}})_1 dA^z\right)^z =$$

$$= -\oint_S \left(b_p \times ds\right)^{\bar{z}} \qquad (1.2.33)$$

The *Stokes' Formula* in the space has the form

$$\int_{V^e} \left((curl\, b_p)^a dV^e\right)_p = \int_{V^e} \left((b_p \times \nabla)^a dV^e\right)_p =$$

$$= \int_{V^e} \left((b_{p,\times\bar{r}})^a dV^e\right)_{\bar{p}} = -\int_A \left(b_p \otimes dA^n\right)_p \qquad (1.2.34)$$

and in the $x, y$-plane, the form

$$\int_{A^z} \left(\left(curl\ b_p\right)^z \cdot dA^z\right)_{\bar{1}} = \int_{A^z} \left(\left(b_p \times {}^z\nabla\right)^z dA^z\right)_{\bar{1}} =$$

$$= \int_{A^z} \left(\left(b_{p,\times\bar{r}}\right)^z dA^z\right)_{\bar{1}} = -\oint_S \left(b_p \cdot ds\right)_{\bar{1}} \qquad (1.2.35)$$

# 2. THE DEFORMABLE CONTINUUM

## 2.1. The Concept of the Deformable Continuum

A deformable continuum is an aggregate of numerous material particles, namely atoms and molecules interconnected through chemical and crystal bonds of electric nature, such that the treatment of the dynamic behavior of the individual particles is impossible and the behavior of the aggregate is described through statistical entities: the number of particles per unit volume, the mass density, the energy density, the pressure, the temperature etc. The values of these statistical entities are considered as continuously varying through the points of the continuum and with the time. Consequently a continuum is defined by a set of various fields taking these values at the different points and for different time instants. It is mentioned that the size of the atoms and molecules varies between $10^{-5}$ and $10^{-8}$ cm, while the size of the continuum media, the solids and the fluids, liquids or gases, contained in the subject of the continuum mechanics exceeds the $10^{-1}$ cm.

As *material points*, which differ from the geometrical points of the space occupied by the continuum, are considered infinitesimal volume segments which in an orthogonal system of reference are represented by the symbol $dV^e = ((dx\, dy\,)^z dz)^e$. Although a sufficiently small number of individual particles is assumed to be contained in the infinitecimal volume of each one material point the occupy average values of their spacings, their velocities and their kinetic energies associated with quasi constant of standard deviation, insuring that they can be considered as homogeneous. In cases where the existence of such a set of particles is not possible, as e.g. in a sparse gas, the continuum approximation is not acceptable.

The motion of the geometrical points consists of translations in the 3-dimensional space represented by polar vectors which are functions of the position and the time. The variations of the translations along infinitesimal lengths of the material points result to their deformation consisting in normal dilatations in the three directions, extensions or contractions, as well as in relative shear sliding between the opposite sides which transform the infinitesimal orthogonal parallelepiped to a skew one. The continuity condition of the medium implies that the interfaces between adjacent material points remain in contact during the deformation. Except the rotations which are associated with the shear deformation, no additional intrinsic rotations of the material points are provided. As a result the translations and the orbital velocities are the main parameters of the motion and the work and the kinetic energy performed are attributed exclusively to the action of forces and orbital momentums, applied on the material points, both represented by polar vectors, while the action of intrinsic moments or angular momentums, represented by axial vectors, is not considered. Nevertheless continuum models providing intrinsic rotations and moments have been proposed in the past, as for example the quasi-elastic model for the ether of Mac Cullagh (1839), but it failed to contribute in the final formulation of Continuum Mechanics for the reasons explained in detail in the next chapter 6.

Continuous media performing motion with finite velocities, as in the case of the flow of fluids, appear in the following four different states. The tranquil, the laminar, the turbulent and the vortical. The later states are actually metastable modes of the motion which are the result of instability conditions occurring at a previous state. For the tranquil and the laminar state the action consisting exclusively of polar vector forces and orbital momentums is obvious. In chapter 6 will be shown that the turbulent and the vortical states can be considered as aggregates of particles of non-infinitesimal size with motion parameters containing rotations and angular velocities under the action of axial vector moments and angular momentums which however are the resultants of polar vector forces and orbital momentums applied on their constituent material points of infinitesimal volume.
The scope of the present chapter is restricted to the presentation in octonion notation of expressions and equations already available from the existing literature on Continuum Mechanics, where they are presented in the traditional algebraic notation.

## 2.2. Translations and Strains

The motion of a deformable continuum is defined by the polar vector field of the translations up of the geometrical points

$$r = \sum_i r_i = x + y + z$$

which are functions of their coordinates and the time

$$u_p[t,x,y,z] = \sum_i u_i[t,x,y,z]$$

(where $i = x, y, z$), with total differential

$$du_p = \sum_i du_i = \sum u_{i,\bar{t}}dt + \sum_i \sum_j u_{i,\bar{j}}dr_j \qquad (2.1)$$

The motion of a *material point* occupying an infinitesimal volume element $d_oV^e = \left(\left(d_{ox}d_{oy}\right)^z d_{oz}\right)^e$ in the viscinity of a geometrical point is defined by the following parameters.

- The rigid body translation

$$u_{p,\bar{t}}dt = \sum u_{i,\bar{t}}dt \qquad (2.2a)$$

- The rigid body rotations, which in the case of infinitesimal deformations are defined by the expressions

$$\left(\phi_{i\bar{j}}\right)^{\bar{k}} = \frac{1}{2}\left(\left(u_{i,\bar{j}}\right)^{\bar{k}} - \theta_{ij}\left(u_{j,\bar{i}}\right)^{k}\right) = \left(\phi_{j\bar{i}}\right)^{k} \qquad (2.2b)$$

implying $\phi_{i\bar{i}} = 0$ for $i = j$. The rotations $\left(\phi_{i\bar{j}}\right)^{\bar{k}}$ are the elements of a 3x3-dimensional symmetric tensor of rank 2 with zero diagonal elements.

- The strains denoting the infinitesimal deformation of the material point, with are defined by the expressions

$$\left(\varepsilon_{i\bar{j}}\right)^{\bar{k}} = \frac{1}{2}\left(\left(u_{i,\bar{j}}\right)^{\bar{k}} + \theta_{i\bar{j}}\left(u_{j,\bar{i}}\right)^{k}\right) = -\left(e_{j\bar{i}}\right)^{k} \qquad (2.2c)$$

The strains $\left(\varepsilon_{i\bar{j}}\right)^{\bar{k}}$ are the elements of a 3x3-dimensional antisymmetric tensor of rank 2. For $i = j$ the diagonal elements are the *normal strains* are $(\varepsilon_{i\bar{i}})_1 = \left(u_{i,\bar{i}}\right)_1$ and represent the dilatation, extension or contraction, per unit length along the direction $r_i$, while for $i \neq j$ the non diagonal elements $\left(\varepsilon_{i\bar{j}}\right)^{\bar{k}}$ are shear strains and represent the relative sliding in the direction $r_i$ transversal to the direction $r_j$ and per unit length of $r_j$.

In (Fig. 2.1) an illustration is presented of the above motion parameters projected on the side $dr_i dr_j$ of the material point.

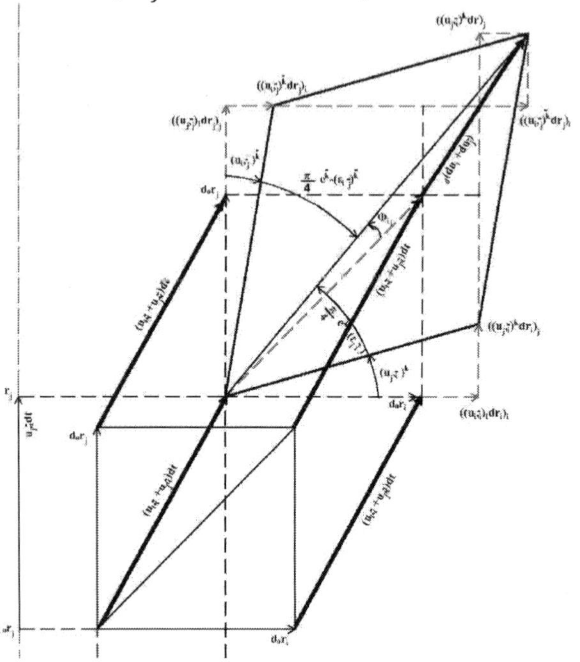

*Figure 2.1 - Translations due to deformation*

The length variations at the edges of the material points are

$$dr_i - d_o r_i = du_i - u_{i,\bar{t}} dt = \sum_j u_{i,\bar{j}} dr_j = \sum_j \left(\varepsilon_{i\bar{j}} + \phi_{i\bar{j}}\right) dr_j \qquad (2.3)$$

and the variation of the length

$$ds = dx + dy + dz = \sum_i dr_i$$

is

$$ds^2 - d_o s^2 = -\sum_i \sum_j \left(\left(\varepsilon_{i\bar{j}}\right)^{\bar{k}} \left(dr_i dr_j\right)^k\right)_1 \qquad (2.4)$$

By the orthogonal transformation

$$\varepsilon_{i'\bar{j}'} = \sum_k \sum_l c_{i'\cdot\bar{k}} \varepsilon_{k\bar{l}} c_{l\cdot\bar{j}'} , \qquad \phi_{i'\bar{j}'} = \sum_k \sum_l c_{i'\cdot\bar{k}} \phi_{k\bar{l}} c_{l\cdot\bar{j}'} \tag{2.5}$$

the motion parameters are obtained in any other orthogonal reference system $(x', y', z')$ defined by the direction cosines $c_{i'\cdot\bar{k}}$ of the unit vectors $e_{i'}$ with respect to the unit vectors $e_k$. Under orthogonal transformations the following entities remain invariant:

$$_1E = \sum_i \varepsilon_{i\bar{i}}, \qquad _2E = \sum_i \|\varepsilon_{i\bar{i}}\|, \qquad _3E = \left\|\left[\varepsilon_{i\bar{j}}\right]\right\| \tag{2.6}$$

where $\left\|\left[\varepsilon_{i\bar{j}}\right]\right\|$ is the determinant of the aray $\varepsilon_{i\bar{j}}$ and

$$\|\varepsilon_{i\bar{i}}\| = \left\|\begin{bmatrix} \varepsilon_{j\bar{j}} & \varepsilon_{j\bar{k}} \\ \varepsilon_{k\bar{j}} & \varepsilon_{k\bar{k}} \end{bmatrix}\right\|$$

are the minor determinants corresponding to the diagonal terms $\varepsilon_{i\bar{i}}$ .

An orthogonal reference system $(r_\zeta, r_n, r_\xi)$ exists where the non diagonal terms $\varepsilon_{\zeta\bar{n}}$ for $n \neq \zeta$, which represent the shear strains, have zero values. The non zero diagonal terms $\varepsilon_{\zeta\bar{\zeta}}, \varepsilon_{n\bar{n}}$ and $\varepsilon_{\xi\bar{\xi}}$ are the *principal normal strains* and are the roots of the algebraic equation

$$\varepsilon_{\zeta\bar{\zeta}}^3 - {_1E}\,\varepsilon_{\zeta\bar{\zeta}}^2 + {_2E}\,\varepsilon_{\zeta\bar{\zeta}} - {^3E} = 0 \tag{2.7}$$

The principal normal strains satisfy the relations

$$\varepsilon_{\zeta\bar{\zeta}} + \varepsilon_{\eta\bar{\eta}} + \varepsilon_{\xi\bar{\xi}} = {_1E} = \sum_i \varepsilon_{i\bar{i}} \tag{2.8$a$}$$

$$\varepsilon_{\zeta\bar{\zeta}}\varepsilon_{n\bar{n}} + \varepsilon_{\eta\bar{\eta}}\varepsilon_{\xi\bar{\xi}} + \varepsilon_{\xi\bar{\xi}}\varepsilon_{\zeta\bar{\zeta}} = {_2E} \tag{2.8b}$$

$$\varepsilon_{\zeta\bar{\zeta}}\varepsilon_{\eta\bar{\eta}}\varepsilon_{\xi\bar{\xi}} = {_3E} \tag{2.8c}$$

The *mean normal strain* is defined by the relation

$$ {}_a\varepsilon_1 = \frac{1}{3}\sum_i (\varepsilon_{i\bar{\imath}})_1 = \frac{{}_1E}{3} \tag{2.9}$$

and the normal strains can be considered as the sum of a *conformal* or *isotropic* one ${}_c\varepsilon_{i\bar{\imath}} = \delta_{ij}\ {}_a\varepsilon_1$ and a *contra-formal* or *diviatoric* one ${}^c\varepsilon_{i\bar{\imath}} = \varepsilon_{i\bar{\imath}} - \delta_{ij}\ {}_a\varepsilon_1$. Consequently the strain tensor is the sum of a conformal or isotropic component with elements

$$ {}_c\varepsilon_{i\bar{\jmath}} = \delta_{ij}\ {}_a\varepsilon_1 \tag{2.10a}$$

and a contra-formal or deviatoric component with elements

$$ {}^c\varepsilon_{i\bar{\jmath}} = \varepsilon_{ij} - \delta_{ij}\ {}_a\varepsilon_1 \tag{2.10b}$$

The volumetric strain ${}_u\varepsilon_1$ defines the relative variation of the volume $d_oV^e = \left(d_or_\zeta d_or_n\right)^{\xi} d_or_\xi$ of the material points and is represented by the expression

$$ \begin{aligned} {}_u\varepsilon_1 &= (dV^e - d_oV^e)\widehat{d_oV^e} = \\ &= \left(\left((1+\varepsilon_{\zeta\bar{\zeta}})(1+\varepsilon_{n\bar{n}})(1+\varepsilon_{\xi\bar{\xi}}) - 1\right) d_oV^e\right)\widehat{d_oV^e} = \\ &= {}_1E + {}_2E + {}_3E \end{aligned} \tag{2.11}$$

For the elastic deformations of stiff solids or of liquids in the tranquil state, which are usually considered as infinitesimals, ${}_2E$ and ${}_3E$ are higher order infinitesimals and are omitted in Eq. (2.9), which in that case takes the simplified approximate expression

$$ {}_V\varepsilon_1 \cong {}_1E = \sum_i (\varepsilon_{i\bar{\imath}})_1 = 3\,{}_a\varepsilon_1 \tag{2.11a}$$

Between the second derivatives of the strains with respect to the coordinates the following *Compatibility Conditions* (Saint-Venant) are satisfied, consisting of the set of three differential equations

$$\left(\varepsilon_{i\bar{j},\bar{k}} + \varepsilon_{j\bar{k},\bar{i}} + \varepsilon_{k\bar{i},\bar{j}}\right)_{,\bar{i}} = \varepsilon_{i\bar{i},\bar{j}k} \tag{2.12a}$$

each one corresponding to a cyclic permutation $(i, j, k)$ of the indices $x, y, z$ or the equivalent set of three equations

$$\varepsilon_{i\bar{i},\bar{j}\bar{j}} + \varepsilon_{j\bar{j},\bar{i}\bar{i}} = \varepsilon_{i\bar{j},\bar{j}\bar{i}} + \varepsilon_{j\bar{i},\bar{i}\bar{j}} \tag{2.12b}$$

each one corresponding to one of the three pairs $i, j$ of the indices.
After applying the compatibility conditions (2.12a or b) the six strain functions $\varepsilon_{i\bar{j}}$ are reduced to three and the estimation of the three translation functions $u_i$ from the Eqs. (2.2c) becomes accessible.

## 2.3 Orbital Velocities and Strain Rates

The motion parameters presented in the previous section 2.2 concern the case of infinitesimal deformations of the continuum which occurs for the elastic deformations of stiff solids or of liquids in the tranquil state at a quasi-static condition. In the cases of plastic behavior of ductile solids and of the flow of liquids and gases the deformations are large and the Eqs. (2.2.b, c) take the extended form

$$\left(\phi_{i\bar{j}}\right)^{\bar{k}} = \frac{1}{2}\left(\left(u_{i,\bar{j}}\right)^{\bar{k}} - \theta_{ij}\left(u_{j,\bar{i}}\right)^{k} \mp \sum\left(u_{k,\bar{i}}\right)^{\bar{j}}\left(u_{k,\bar{j}}\right)^{i}\right)^{\bar{k}}$$
$$= \left(\phi_{j\bar{i}}\right)^{k} \tag{2.13a}$$
$$\left(\varepsilon_{i\bar{j}}\right)^{\bar{k}} = \frac{1}{2}\left(\left(u_{i,\bar{j}}\right)^{\bar{k}} - \theta_{ij}\left(u_{j,\bar{i}}\right)^{k} \pm \sum_{k}\left(u_{k,\bar{i}}\right)^{\bar{j}}\left(u_{k,\bar{j}}\right)^{i}\right)^{\bar{k}}$$
$$= -\left(\varepsilon_{j\bar{i}}\right)^{k} \tag{2.13b}$$

In Eqs. (2.13a, b) the upper sign is applied for the Lagrange's procedure, where variables are the initial coordinates $_0x$, $_0y$, $_0z$ of the geometrical points, while the lower sign is applied for the Euler's procedure, where variables are the final coordinates $x, y, z$.

The expressions (2.13a,b) imply the following difficulties. They constitute non-linear differential equations resulting to complicated solutions.

Moreover the finite rigid rotations are not commutative in addition and the result of their superposition depends on the sequence of their development, although they are developed simultaneously around the three axes of rotation considered. All these difficulties are canceled in the usual case where large translations and deformations are developed for continuous media executing motions with finite velocities of their material points, as it is for example the flow of the fluids or the plastic deformation of ductile solids. By considering as basic motion parameter the orbital velocities $u_p = du_p\bar{d}t$ instead of the translations $u_p$, as well as the rigid body angular velocities $\omega_{i\bar{j}} = \phi_{i\bar{j},\bar{t}}$ instead of the rotations $\phi_{i\bar{j}}$ and the strain rates $\gamma_{i\bar{j}} = \varepsilon_{i\bar{j},\bar{t}}$ instead of the strains $\varepsilon_{i\bar{j}}$ in cases infinitesimal displacements of the geometrical points are executed at infinitesimal time intervals, relations similar to those of the section 2.2 are obtained in the following manner.

The orbital velocity of a geometrical point, after taking into account Eq. (2.1), will be

$$v_p = du_p\widehat{dt} = \sum_i du_i\widehat{dt} = \sum_i u_{i,\bar{t}} + \sum_i\sum_j u_{i,\bar{j}}dr_j\widehat{dt} =$$
$$= \sum_i u_{i,\bar{t}} + \sum_i\sum_j \left(u_{i,\bar{t}}\right)_{,\bar{j}} dr_j = \sum_i u_i + \sum_i\sum_j u_{i,\bar{j}}dr_j \tag{2.14}$$

The orbital acceleration $a_p$ will be

$$a_p = du_p\widehat{dt} = \sum_i v_{i,t} + \sum_i\sum_j v_{i,\bar{j}}\left(dr_j\widehat{dt}\right) =$$
$$= \sum_i v_{i,\bar{t}} + \sum_i\sum_j \left(v_{i,\bar{j}}v_j\right)_i \tag{2.15}$$

In Eq. (2.15) $v_{i,\bar{t}}$ represents the *local acceleration*, while

$$\sum_j \left(v_{i,\bar{j}}v_j\right)_i$$

represents the *material acceleration*, both in the direction $r_{\bar{i}}$. For steady flow conditions the local acceleration is zero.

The motion parameters of the material points, according to Eq. (2.14), are:

- The rigid body orbital velocities

$$v_p = \sum v_i = \sum u_{i,\bar{t}} \tag{2.16a}$$

- The rigid body angular velocities

$$\left(\omega_{i\bar{j}}\right)^{\bar{k}} = \frac{1}{2}\left(\left(v_{i,\bar{j}}\right)^{\bar{k}} - \theta_{ij}\left(v_{j,\bar{i}}\right)^{k}\right) = \left(\phi_{i\bar{j}}\right)^{\bar{k}}_{,\bar{t}} = \left(\omega_{j\bar{i}}\right)^{k} \tag{2.16b}$$

- The strain rates

$$\left(\gamma_{i\bar{j}}\right)^{\bar{k}} = \frac{1}{2}\left(\left(v_{i,\bar{j}}\right)^{\bar{k}} - \theta_{ij}\left(v_{j,\bar{i}}\right)^{k}\right) = \left(\varepsilon_{i\bar{j}}\right)^{\bar{k}}_{,\bar{t}} = \left(\gamma_{j\bar{i}}\right)^{k} \tag{2.16c}$$

Additional relations similar to those obtained in section 2.2 result by simple replacement of the strains $\varepsilon_{i\bar{j}}$ with the strain rates $\gamma_{i\bar{j}}$. In correspondence with Eqs. (2.6) to (2.12) the following entities are defined.

The strain rate invariants under orthogonal transformation

$${}_1\Gamma = \sum_i \gamma_{i\bar{i}}, \qquad {}_2\Gamma = \sum_i \|\gamma_{i\bar{i}}\|, \qquad {}_3\Gamma = \left\|\left[\gamma_{i\bar{j}}\right]\right\| \tag{2.17}$$

The principal normal strain rates as roots of the algebraic equation

$$\gamma_{\zeta\bar{\zeta}}^3 - {}_1\Gamma\gamma_{\zeta\bar{\zeta}}^2 + {}_2\Gamma\gamma_{\zeta\bar{\zeta}} - {}_3\Gamma = 0 \tag{2.18}$$

The mean normal strain rate

$$_a\gamma_1 = \frac{1}{3}\sum_i (\gamma_{i\bar{\imath}})_1 = \frac{_1\Gamma}{3} \tag{2.19}$$

The elements of the conformal or isotropic component

$$_c\gamma_{i\bar{\jmath}} = \delta_{ij}\ _a\gamma_1 \tag{2.20a}$$

and those of the contra-formal or deviatoric component

$$^c\gamma_{i\bar{\jmath}} = \gamma_{i\bar{\jmath}} - \delta_{ij}\ _a\gamma_1 \tag{2.20b}$$

of the strain rate tensor.

The volumetric strain rate

$$_V\gamma_1 = \ _1\Gamma + \ _2\Gamma + \ _3\Gamma \cong \ _1\Gamma = 3\ _a\gamma_1 \tag{2.21}$$

At last the following compatibility conditions are satisfied

$$\left(\gamma_{i\bar{\jmath},\bar{k}} + \gamma_{j\bar{k},\bar{\imath}} + \gamma_{k\bar{\imath},\bar{\jmath}}\right)_{,\bar{l}} = \gamma_{i\bar{l},\bar{\jmath}\bar{k}} \tag{2.22a}$$

or the equivalent

$$\gamma_{i\bar{\imath},j\bar{\jmath}} + \gamma_{j\bar{\jmath},i\bar{\imath}} = \gamma_{i\bar{\jmath},j\bar{\imath}} + \gamma_{j\bar{\imath},i\bar{\jmath}} \tag{2.22b}$$

## 2.4 Forces and Stresses

On the material points of a deformable continuum the following kind of polar vector forces are applied, which are functions of the position vector $r$ of the geometrical points.

The *inertial forces*

$$dg_p = \sum_i dg_i = g_p{}^{\bar{e}} dV^e = \sum_i g_i{}^{\bar{e}} dV^e = a_p \rho_1{}^{\bar{e}} dV^e \tag{2.23}$$

where : $\rho_{1\bar{e}}$ is the mass density of the material

$a_p = \sum_i a_i$ is the acceleration induced by external potential fields

$g_{p\bar{e}} = \sum_i g_{i\bar{e}}$ is the inertial forces per unit volume

The *external surface loads* applied on infinitesimal boundary surface elements $dA^n$ of the continuum segment with unit normal vector $e_n$

$$dq_p = \sum_i dq_i = \sum_i (q_i{}^{\bar{n}})\, dA^n \qquad (2.24)$$

where $q_i{}^{\bar{n}}$ are the external forces per unit surface $A^n$.

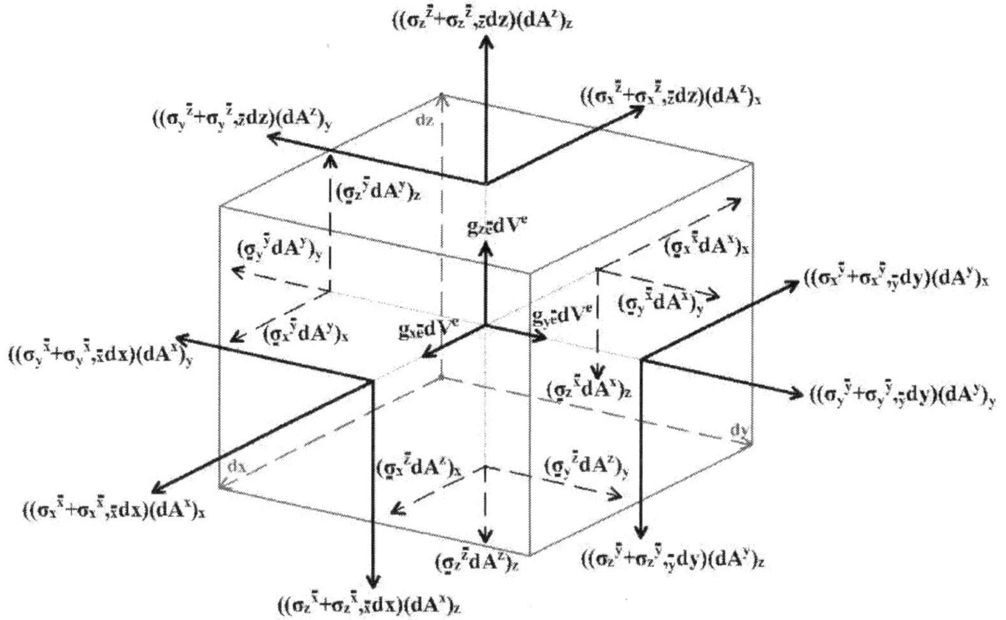

*Figure 2.2 - Stresses on a cubic element*

The *stresses* which represent the interacting forces between adjacent material points and are developed on their interfaces considered with their values per unit area of them (Fig. 2.2). The stresses developed on the interface $dA^j = (dr_k dr_i)^j$ lying at a position with coordinate $r_j$ are

$$ {}^{j}d\underline{\sigma}_{p} = \sum_{i} {}^{j}d\underline{\sigma}_{i} = \sum_{i} \underline{\sigma}_{i}{}^{\bar{j}} dA^{j} \qquad (2.25a) $$

and those developed on the parallel interface $dA^j$ lying at the neighboring position $r_j + dr_j$ are

$$ {}^{j}(d\sigma_{p} + \Delta d\sigma_{p}) = \sum_{i} {}^{j}(d\sigma_{i} + \Delta d\sigma_{i}) = \sum_{i} \left(\sigma_{i}{}^{\bar{j}} + \sigma_{i}{}^{\bar{j}}{}_{,\bar{j}} dr_{j}\right) dA^{j} \qquad (2.25b) $$

The stresses $\sigma_i{}^{\bar{j}}$ arranged at an array of 3x3 dimensions are elements of the *stress tensor* of rank 2. The diagonal elements $\left(\sigma_j{}^{\bar{j}}\right)_e$ of the tensor, corresponding to $i = j$, are the normal stresses and represent interacting forces acting in the direction normal to the plane of the interface $dA^j$. The non diagonal elements $(\sigma_i{}^{\bar{j}})_{\bar{k}}$, corresponding to $i \neq j$, are the *shear stresses* and represent interacting forces with directions parallel to the plane of the interface $dA^j$.

By the orthogonal transformation

$$ \sigma_{i'}{}^{\bar{j}'} = \sum_{k} \sum_{l} c_{i' \cdot \bar{k}} \sigma_{k}{}^{\bar{l}} c^{l \cdot \bar{j}'} \qquad (2.26) $$

the stresses are obtained in any other orthogonal reference system ($x'$, $y'$, $z'$) defined by the direction cosines $c_{i'\cdot\bar{k}}$ and $c^{l\cdot\bar{j}'} = c_{l\cdot\bar{j}'}$ of the unit vectors $e_{i'}$with respect to the unit vectors $e_k$. Under orthogonal transformations the following entities remain invariant:

$$ {}_{1}S = \sum_{i} \sigma_{i}{}^{\bar{i}}, \qquad {}_{2}S = \sum_{i} \|\sigma_{i}{}^{\bar{i}}\|, \qquad {}_{3}S = \|[\sigma_{i}{}^{\bar{j}}]\| \qquad (2.27) $$

where $\|[\sigma_i{}^{\bar{j}}]\|$ is the determinant of the aray $\sigma_i{}^{\bar{j}}$ and

$$ \|\sigma_{i}{}^{\bar{i}}\| = \left\| \begin{bmatrix} \sigma_{j}{}^{\bar{j}} & \sigma_{j}{}^{\bar{k}} \\ \sigma_{k}{}^{\bar{j}} & \sigma_{k}{}^{\bar{k}} \end{bmatrix} \right\| $$

are the minor determinants corresponding to the diagonal terms $\sigma_i{}^{\bar{i}}$.

An orthogonal reference system $(r_\zeta, r_\eta, r_\xi)$ exists where the non diagonal terms $\sigma_\zeta{}^{\bar{\eta}}$ for $\eta \neq \zeta$, which represent the shear stresses, have zero values. The non-zero diagonal terms $\sigma_\zeta{}^{\bar{\zeta}}, \sigma_\eta{}^{\bar{n}}$ and $\sigma_\xi{}^{\bar{\xi}}$ are the *principal normal stresses* and are the roots of the algebraic equation

$$\sigma_\zeta{}^{\bar{\zeta}3} - {}_1S\sigma_\zeta{}^{\bar{\zeta}2} + {}_2S\sigma_\zeta{}^{\bar{\zeta}} - {}_3S = 0 \tag{2.28}$$

The principal normal stresses satisfy the relations

$$\sigma_\zeta{}^{\bar{\zeta}} + \sigma_n{}^{\bar{n}} + \sigma_\xi{}^{\bar{\xi}} = {}_1S = \sum_i \sigma_i{}^{\bar{\iota}} \tag{2.29a}$$

$$\sigma_\zeta{}^{\bar{\zeta}} + \sigma_\eta{}^{\bar{\eta}} + \sigma_\eta{}^{\bar{\eta}}\sigma_\xi{}^{\bar{\xi}} + \sigma_\xi{}^{\bar{\xi}}\sigma_\zeta{}^{\bar{\zeta}} = {}_2S \tag{2.29b}$$

$$\sigma_\zeta{}^{\bar{\zeta}}\sigma_\eta{}^{\bar{\eta}}\sigma_\xi{}^{\bar{\xi}} = {}_3S \tag{2.29c}$$

The *mean normal* (or *hydrostatic*) stress is defined by the relation

$${}_a\sigma^e = \frac{1}{3}\sum_i (\sigma_i{}^{\bar{\iota}})_e = \frac{{}_1S}{3} \tag{2.30}$$

Each one normal stress can be considered as the sum of a *conformal* or *isotropic* component ${}_c\sigma_i{}^{\bar{\iota}} = {}_a\sigma^e$ and a *contra-formal* or *deviatoric* one ${}^c\sigma_i{}^{\bar{\iota}} = \sigma_i{}^{\bar{\iota}} - {}_a\sigma^e$.

Consequently the stress tensor is the sum of a conformal or isotropic component with elements

$${}_c\sigma_i{}^{\bar{J}} = \delta_{ij}\ {}_a\sigma^e \tag{2.31a}$$

and a contra-formal or deviatoric component with elements

$${}^c\sigma_i{}^{\bar{J}} = \sigma_i{}^{\bar{J}} - \delta_{ij}\ {}_a\sigma^e \tag{2.31b}$$

The following six *equilibrium conditions* between the stresses and the inertial forces applied on a material point result from Eqs. (2.25a, b) and the (Fig. 2.2):

$$\underline{\sigma_i}^{\bar{\imath}}{}_{,\bar{\imath}} + \sigma_i{}^{\bar{J}}{}_{,\bar{J}} + \sigma_i{}^{k}{}_{,\bar{k}} + g_i{}^{\bar{e}} = 0 \tag{2.32a}$$

and

$$\sigma_i{}^{\bar{J}} = -\sigma_j{}^{\bar{\imath}} \; (for \; i \neq j), \qquad (1-\delta_{ij})\sigma_i{}^{\bar{J}} = -(1-\delta_{ji})\sigma_j{}^{\bar{\imath}} \tag{2.32b}$$

Eqs. (2.32a) result from the equilibrium between the polar vector forces, while Eqs. (2.32b) result from the equilibrium between the axial vector moments of the forces with respect to the center of mass of the material point. The moments of the forces $\left(\sigma_i{}^{\bar{J}}{}_{,\bar{J}} dj\right) dA^j$ are higher order infinitecimals and their contribution is omitted. Eq. (2.32b) implies that the stress tensor as well as its deviatoric component are antisymmetric.

Between the external surface loads $q_i{}^{\bar{\eta}}$ and the stresses applied on material points with one face $dA^n$ lying on the boundary surface of the continuum segment considered, the following three boundary conditions are satisfied (Fig. 2.3)

$$p_i{}^{\bar{n}} = \sum_j \sigma_i{}^{\bar{J}} \, c_{j \cdot \bar{n}} \tag{2.33}$$

where $c_{j \cdot \bar{n}} = c^{j \cdot \bar{n}} = dA^j \cdot dA^n$ are the direction cosines of the unit vectors $e_j$ with respect to the unit vector $e^n$ normal to the infinitecimal are $adA^n$ of the face of the material point on the boundary surface. The inertial forces $g_{i\bar{e}} dV^e$ and the stress resultants $\left(\sigma_i{}^{\bar{J}}{}_{,\bar{J}} dj\right) dA^j$ are higher order infinitecimals and their contribution is omitted.

## 2.5 Orbital Momentums and Impulses

The potential energy associated with motions and deformations of the material points with infinitesimal velocities, which correspond to the quasi-static state of the continuum, consists exclusively to minus the work performed by the forces and strains presented in the previous section 2.4 for the corresponding translations and strains presented in the section 2.2. For motions with finite velocities, as for example in the case of the flow of fluids, to the total energy the kinetic energy have to be added, represented by half the sum of the products of the orbital momentums and impulses by the velocities and strain rates respectively.

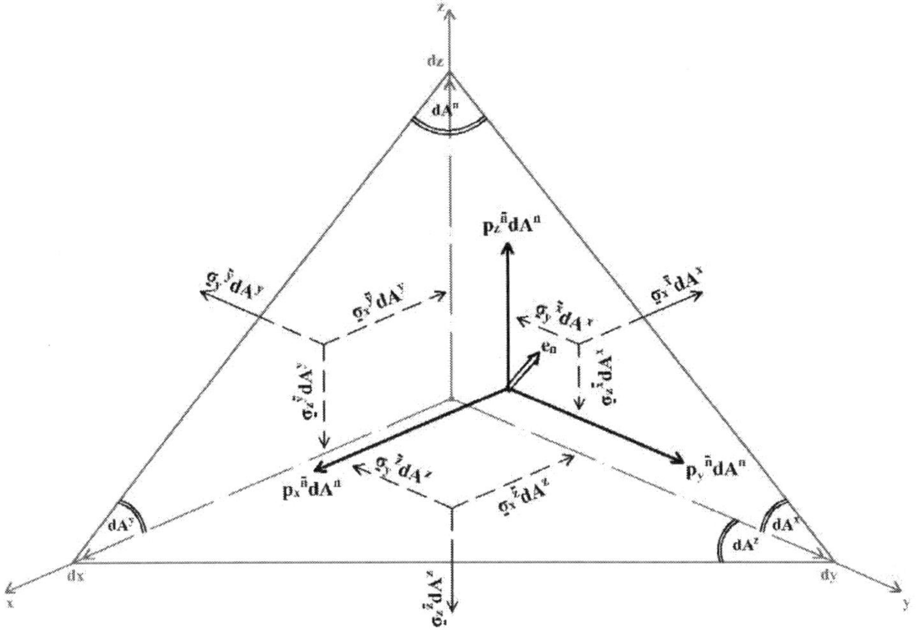


*Figure 2.3 - Stresses on a boundary surface*

The inertial forces as well as the external surface loads and the stress resultants are the derivatives with respect to the time of the orbital momentums $p_p$ as well as of the *shots* applied on the external boundary surface $s_i{}^{\bar{n}}dA^n$ and the *corpuscles* developed on the interfaces between the material points $\tau_i{}^{\hat{j}}dA^j$ respectively, where as shots $s_i{}^{\hat{n}}$ and compulses $\tau_i{}^{\hat{j}}$ are meant impulses per unit area.

For the mass orbital momentum $dp_p = \upsilon_p dm_1$

$$dg_p = \sum_i g_{i\bar{e}} dV^e = \sum_i p_{i\bar{e},t}\, dV^e = dp_{p,t} \qquad (2.34a)$$

For the external shots

$$dq_p = \sum_i q_i{}^{\bar{n}} dA^n = \sum_i s_i{}^{\bar{n}}{}_{,\bar{t}} dA^n = ds_{p,\bar{t}} \qquad (2.34b)$$

For the internal corpuscles

$$ {}^{j}d\sigma_{p} = \sum_{i} \sigma_{i}{}^{\bar{J}} dA^{j} = \sum_{i} \tau_{i}{}^{\bar{J}}{}_{,\bar{t}} dA^{j} = {}^{j}d\,\tau_{p,\bar{t}} \tag{2.34c} $$

Additional relations similar to those obtained in section 2.4 result by simple replacement of the stresses $\sigma_{i}{}^{\bar{J}}$ with the corpuscles $\tau_{i}{}^{\bar{J}}$. In correspondence with Eqs. (2.27) to (2.33) the following entities are defined.

The internal corpuscle invariants under orthogonal transformation

$$ {}_{1}\mathrm{T} = \sum_{i} \tau_{i}{}^{\bar{\imath}}, \quad {}_{2}\mathrm{T} = \sum_{i} ||\tau_{i}{}^{\bar{\imath}}||, \quad {}_{3}\mathrm{T} = ||[\tau_{i}{}^{\bar{J}}]|| \tag{2.35} $$

The principal normal corpuscles as roots of the algebraic equation

$$ \left(\iota_{\zeta}{}^{\bar{\zeta}}\right)^{3} - {}_{1}\mathrm{T}\left(\tau_{\zeta}{}^{\bar{\zeta}}\right)^{2} + {}_{2}\mathrm{T}\tau_{\zeta}{}^{\bar{\zeta}} - {}_{3}\mathrm{T} = 0 \tag{2.36} $$

The mean normal corpuscle

$$ {}_{a}\tau^{e} = \frac{1}{3}\sum (\tau_{i}{}^{\bar{\imath}})^{e} = \frac{{}_{1}T}{3} \tag{2.37} $$

The elements of the conformal or isotropic component

$$ {}_{c}\tau_{i}{}^{\bar{J}} = \delta_{ij\,a}\tau^{e} \tag{2.38a} $$

and those of the contra-formal or deviatoric component of the corpuscle tensor.

$$ {}^{c}\tau_{i}{}^{\bar{J}} = \tau_{i}{}^{j} - \delta_{ij\,a}\tau^{e} \tag{2.38b} $$

At last, the equilibrium conditions

$$ \tau_{i}{}^{\bar{\imath}}{}_{,\bar{\imath}} + \tau_{i}{}^{\bar{J}}{}_{,\bar{J}} + \tau_{i}{}^{\bar{k}}{}_{,\bar{k}} + p_{i}{}^{\bar{e}} = 0 \tag{2.39} $$

and the boundary conditions are satisfied.

$$ s_{i}{}^{\bar{n}} = \sum_{j} \tau_{i}{}^{\bar{J}}\, c_{j\cdot\bar{n}} \tag{2.40} $$

## 2.6 Constitutive Laws of Solids

The material property affecting the relations between the dynamic entities of forces and momentums, due to potential or inertial effects and the kinematic entities of velocities and accelerations, is the mass. The mass $d_{m_1} = {\rho_1}^{\bar{e}} dV^e$ can be defined as the force required for a material point to acquire a unit acceleration, or, alternatively, the momentum required the material point to acquire a unit velocity, while the mass density ${\rho_1}^{\bar{e}}$ is a characteristic constant of the material of the continuum which is a function of the position vector $r$.

The relations between stresses and strains or between corpuscles and strain rates, which are developed at the interfaces between material points where volume and mass disappear, depend on characteristic constants representing stiffness or deformability properties of the material and are referred as the *constitutive laws* of the material.

For solid elastic materials the Hooke's constitutive law is applied, which is represented by the nine linear equations, corresponding to the nine stresses

$$({\sigma_i}^{\bar{J}})_{\bar{k}} = \sum_l \sum_m E(i^{\bar{J}})_{\bar{k}(\overline{lm})^n} (\varepsilon_{l\bar{m}})^{\bar{n}} \qquad (2.41)$$

The two groups of indices $i, j, k$ and $l, m, n$ have to be considered as cyclic permutations of the indices $x, y, z$. The symbols $E$ followed by the two groups of indices $(i^{\bar{J}})_{\bar{k}}$ and $(lm)^{\bar{n}}$ represent the elastic constants of the material, expressed in $(Nm^{-2})$ in the MKS system, and constitute the elements of a tensor of rank 4 represented by the 9X9-dimensional matrix:

| | $(\varepsilon_{l\bar{l}})_1$ 1x3 | $(\varepsilon_{l\bar{m}})^{\bar{n}}$ 1x3 | $(\varepsilon_{m\bar{l}})^n$ 1x3 |
|---|---|---|---|
| $({\sigma_i}^{\bar{\iota}})^e$ 3x1 | $E(i^{\bar{\iota}})^e{}_{(l\bar{l})_1}$ 3x3 | $E(i^{\bar{\iota}})^e{}_{(\overline{lm})^n}$ 3x3 | $E(i^{\bar{\iota}})^e{}_{(\overline{ml})^{\bar{n}}} = -E(i^{\bar{\iota}})^e{}_{(\overline{lm})^n}$ 3x3 |
| $({\sigma_i}^{\bar{J}})_{\bar{k}}$ 3x1 | $E(i^{\bar{J}})_{\bar{k}(l\bar{l})1}$ 3x3 | $E(i^{\bar{J}})_{\bar{k}(\overline{lm})^n}$ 3x3 | $E(i^{\bar{J}})_{\bar{k}(\overline{ml})^{\bar{n}}} = -E(i^{\bar{J}})_{\bar{k}(\overline{lm})^n}$ 3x3 |
| $({\sigma_j}^{\bar{\iota}})_k$ 3x1 | $E(j^{\bar{\iota}})_{k(ll)1} = -E(i^{\bar{J}})_{\bar{k}(l\bar{l})}1$ 3x3 | $E(j^{\bar{\iota}})_{k(\overline{ml})^n} = -E(i^{\bar{J}})_{\bar{k}(\overline{lm})^n}$ 3x3 | $E(j^{\bar{\iota}})_{k(\overline{ml})^{\bar{n}}} = -E(j^{\bar{\iota}})_{k(\overline{lm})^n} = -E(i^{\bar{J}})_{\bar{k}(\overline{ml})^n} = E(i^{j})_{\bar{k}(\overline{lm})^n}$ 3x3 |

From the relations

$$(\sigma_i^{\bar{j}})_{\bar{k}} = -(\sigma_j^{\bar{i}})_{\bar{k}}, \qquad (\varepsilon_{i\bar{j}})^{\bar{k}} = -(\varepsilon_{j\bar{i}})^{k}$$

result the relations shown in the three last rows and the three last columns of the matrix which imply that from the 81 constants the 45 contained in these rows and columns have values equal to those of the rest 36 with opposite sign.

The 36 constants have distinct values in case of adiabatic state where the total heat of the closed system of the material points remains constant during the deformation. In the case of isothermal state, where the temperature at each one material point remains constant during the deformation, there exists an elastic potential $Y_1^{\bar{e}}$ equal to the strain energy per unit volume which satisfies the relation

$$(\sigma_i^{\bar{j}})_{\bar{k}} = \left(\partial Y_1{}^{\bar{e}}\overline{\partial(\varepsilon_{i\bar{j}})^{\bar{k}}}\right)_{\bar{k}} \tag{2.42}$$

By differentiating Eq. (2.41) and replacing from (2.42) results for $i \neq j,\ l \neq m$,

$$\begin{aligned} E(i^{\bar{j}})_{\bar{k}(l\bar{m})^n} &= \partial(\sigma_i^{\bar{j}})_{\bar{k}}\overline{\partial(\varepsilon_{l\bar{m}})^{\bar{n}}} = \left(\partial Y_1{}^{\bar{e}}\overline{\partial(\varepsilon_{i\bar{j}})^{\bar{k}}}\right)_{\bar{k}}\overline{\partial(\varepsilon_{l\bar{m}})^{\bar{n}}} = \\ &= -\left(\partial Y_1{}^{\bar{e}}\overline{\partial(\varepsilon_{l\bar{m}})^{\bar{n}}}\right)_{\bar{n}}\overline{\partial(\varepsilon_{i\bar{j}})^{\bar{k}}} = -\partial(\sigma_l{}^{\bar{m}})_{\bar{n}}\overline{\partial(\varepsilon_{i\bar{j}})^{\bar{k}}} = \\ &= -E(l^{\bar{m}})_{\bar{n}(i\bar{j})^k} \end{aligned} \tag{2.43a}$$

while for $i = j$ and $l = m$ it is

$$E(\,i\ \ ^{\bar{\imath}})^{e}{}_{(l\bar{l})\bar{1}} = \mathrm{E}(\,l\ \ ^{\bar{l}})^{e}{}_{(i\bar{\imath})\bar{1}} \tag{2.43b}$$

From (2.43a,b) results that the matrix of the elastic constants is antisymmetric except the upper left 3X3 submatrix which is symmetric and the number of distinct values of the elastic constants is furthermore reduced from 36 to 21. After integrating Eq. (2.42) and replacing Eq. (2.41) the following expression results for the elastic potential:

$$Y_1^{\bar{e}} = \frac{1}{2}\sum_i \sum_j (\sigma_i^{\bar{J}})_{\bar{k}} (\varepsilon_{i\bar{J}})^{\bar{k}} =$$

$$= \frac{1}{2}\sum_i \sum_j \sum_l \sum_m \left(E(i^{\bar{J}})_{\bar{k}}\overline{(\quad l\bar{m})}^{n} (\varepsilon_{l\bar{m}})^{\bar{n}}\right) (\varepsilon_{i\bar{J}})^{\bar{k}} \qquad (2.44)$$

In the case the deformation is executed with energy dissipation or absorption from the closed system the relations between stresses and strains are not furthermore linear and Eqs. (2.41) are transformed to differential ones by replacing $\sigma_i{}^{\bar{J}}$ and $\varepsilon_{l\bar{m}}$ by their differentials $\partial\sigma_i{}^{\bar{J}}$ and $\partial\varepsilon_{l\bar{m}}$ and the constants $E(i^{\bar{J}})_{\overline{k(l\bar{m})^n}}$ by functions having arguments the set of the strains $\varepsilon_{l\bar{m}}$. In that case the system is not conservative and the deformation is not totally reversible.

The equivalent system of linear equations in terms of strains has the form

$$(\varepsilon_{i\bar{J}})^{\bar{k}} = \sum_l \sum_m Z_{(i\bar{J})^{\bar{k}}}\left(_l{}^{\bar{m}}\right)_n (\sigma_l{}^{\bar{m}})_{\bar{n}} \qquad (2.45)$$

where $Z$ are the flexibility constants, equal to the elements of the inverse matrix of $E$.
The distinct values of the elastic constants imply different stiffness of the material in the different directions and for the different modes of deformation, normal or shear, and characterize the anisotropic materials. Anisotropic is the property of solid materials with the ions, atoms or molecules forming crystals which are arranged with the same orientation of their axes at the nodes of a canonical space lattice and constitute a structure referred as mono-crystal. Solid materials with random orientation of the individual crystals or of mono-crystalic granules forming poly-crystalic aggregates or with totally amorphous arrangement of the constituent particles appear the same stiffness in any direction and are referred as isotropic. In any case the internal energy of their structure is represented by the potential of the field of the attractive forces between the constituents particles and crystals.

For isotropic materials Eqs. (2.41) and (2.45) receive the following simplified expressions:

$$\sigma_i{}^{\bar{\iota}} = C^e\left(\varepsilon_{i\bar{\iota}} + k_1\varepsilon_{j\bar{\jmath}} + k_1\varepsilon_{k\bar{k}}\right) \tag{2.46a}$$
$$\sigma_i{}^{\bar{\jmath}} = 2G^e\varepsilon_{i\bar{\jmath}}, \qquad (i \neq \bar{\jmath}) \tag{2.46b}$$
$$\varepsilon_{i\bar{\iota}} = \hat{E}^e\left(\sigma_i{}^{\bar{\iota}} - \nu_1\sigma_j{}^{\bar{\jmath}} - \nu_1\sigma_k{}^{\bar{k}}\right) \tag{2.46c}$$

From the elastic constants contained in Eqs. (2.46a, b, c) the following two, the Young's modulus $E^e$ and the Poisson's ratio $\nu_1$ are considered as basic while the rest are defined by the relations

the shear modulus $$2G^e = \left(\frac{1}{1+\nu_1}\right)E^e \tag{2.47a}$$

the oedometric modulus $$C^e = \frac{(1-\nu_1)}{(1+\nu_1)(1-2\nu_1)}\,E^e \tag{2.47b}$$

the lateral resistance ratio $$k_1 = \frac{\nu_1}{1-\nu_1} \tag{2.47c}$$

The correspondence of the above constants for isotropic materials with those contained in Eqs. (2.41) and (2.45) is shown by the following relations:

$$E\left({}_i{}^{\bar{\iota}}\right)_{\overline{(\iota\bar{\iota})}} = C^e, \qquad E\left({}_i{}^{\bar{\iota}}\right)_{(l\bar{l})} = C^e k_1 \tag{2.48a}$$
$$Z_{(i\bar{\iota})}\overline{\left({}_\iota{}^{\bar{\iota}}\right)} = \widehat{E^e}, \qquad Z_{(i\bar{\iota})}\overline{\left({}_l{}^{\bar{l}}\right)} = -\widehat{E^e}\nu_1 \tag{2.48b}$$

$$E(i^{\bar{\jmath}})_{\overline{(\iota\bar{\jmath})}} = \bar{Z}_{(i\bar{\jmath})}\overline{(\iota^{\bar{\jmath}})} = \pm 2G^e \tag{2.48c}$$

The constants not contained in Eqs. (2.48a, b, c) are all equal to zero.

The following relation between the hydrostatic stress ${}_a\sigma^e$ and the volumetric strain ${}_u\varepsilon_1$ results from Eqs. (2.46a):

$${}_a\sigma^e = K^e\,{}_V\varepsilon_1, \qquad K^e = \frac{1}{3(1-2\nu_1)}E^e \tag{2.49}$$

where $K^e$ is the bulk modulus. From (2.49) results that 0.5 is the maximum possible value of $\nu_1$, corresponding to materials deformed with constant volume.

For deformations executed with finite, instead of infinitesimal, velocities, as for example during elastic vibrations of solid bodies, the corpuscles $\tau_i{}^{\bar{J}}$ and the strain rates $\gamma_{i\bar{J}}$ developed at the interfaces between the material points, instead of the stresses and the strains, are considered. After differciating Eq. (2.41) with respect to the time $t$ and replacing $\varepsilon_{i\bar{J}}$ and $\sigma_i{}^{\bar{J}}$ by the $\gamma_{i\bar{J}}$ and $\tau_i{}^{\bar{J}}$ from Eqs. (2.16c) and (2.34c) respectively the constitudive law of anisotropic materials expressed in terms of the $\tau_i{}^{\bar{J}}$ and $\gamma_{i\bar{J}}$ receives the following form of second order differential equations:

$$\left(\tau_{i,}{}^{\bar{J}}{}_{\bar{t}\bar{t}}\right) = \sum_{l}\sum_{m} E(i^{\bar{J}})_{(l\bar{m})}\gamma_{l\bar{m}} \tag{2.50}$$

For isotropic material the constitutive law in terms of the $\gamma_{i\bar{J}}$ and $\tau_i{}^{\bar{J}}$, in correspondence with Eqs. (2.46a, b, c) receives the expressions:

$$\tau_i{}^{\bar{\imath}}{}_{,\bar{t}\bar{t}} = C^e\left(\gamma_{ii} + k_1\gamma_{j\bar{J}} + k_1\gamma_{k\bar{k}}\right) \tag{2.51a}$$

$$\tau_i{}^{\bar{J}}{}_{,\bar{t}\bar{t}} = 2G^e\gamma_{i\bar{J}}, \qquad (i \neq j) \tag{2.51b}$$

$$\gamma_{i\bar{\imath}} = E^e\left(\tau_i{}^{\bar{\imath}}{}_{,\bar{t}\bar{t}} - \nu_1\tau_j{}^{\bar{J}}{}_{,\bar{t}\bar{t}} - \nu_1\tau_k{}^{\bar{k}}{}_{,\bar{t}\bar{t}}\right) \tag{2.51c}$$

The constitutive law, represented by the equations (2.41) and (2.50), is satisfied for an interval of values of the stresses, determined by two boundary states of the stress tensor, referred as *yield states*, the one corresponding to a positive (tensile) and the other to a negative (compressive) hydrostatic stress. A state of the stress tensor is considered to represent a yield state when it satisfies a relation referred as a *yielding criterion.* Such a relation proposed for general application is the *extended Drucker-Prager criterion*, represented by the equation

$$a_1S + b\sqrt{{}_2S} + c\sqrt[3]{{}_3S} = 1 \tag{2.52}$$

where ${}_1S$, ${}_2S$ and ${}_3S$ are the first, second and third degree respectively invariants of the stress tensor and $a, b, c$ characteristic constants of the material.

Beyond the two boundary yielding states the variation of the strains continues under a quasi constant state of the stress tensor, in an interval of values determined by two boundary states of the strain tensor, refereed as *fracture states*, the one corresponding to a positive and the other to a negative volumetric strain, for which the continuity of the material is interrupted. The materials for which the yield and the fracture states coincide are characterized as *brittle*, otherwise they are characterized as *ductile*. For slightly increasing hydrostatic stress with increasing volumetric strain in the interval between yielding and fracture the materials are furthermore characterized as *strain hardening* and for slightly decreasing hydrostatic stress as *strain softening*. The ductile deformations in that interval are mainly due to the gradual failure of the bonds between atoms and molecules in the microstructure of the material and is accompanied by dissipation of internal bond energy in the environment in the form of heat. The fracture criteria defining the fracture states are expressed in terms of the dissipated internal bond energy.

For isotropic materials the elastic constants are related with the propagation velocity of elastic waves through the material by the Eqs.

$$ {}_{l}c^2 = C^e\overline{\rho_1{}^{\bar{e}}}, \qquad {}_{s}c^2 = G^e\overline{\rho_1{}^{\bar{e}}} \tag{2.53}$$

where ${}_{l}c$: the propagation velocity of longitudinal waves or waves due to dilatations

${}_{s}c$: the propagation velocity of transversal waves or waves due to shear

$\rho_1{}^{\bar{e}}$: the mass density

## 2.7 Constitutive Laws of Liquids

The liquids, considered from the mechanical point of view, are homogeneous isotropic materials performing a ductile behavior between the yielding and the fracture state of the intermolecular bonds, without interruption of the continuity, which permits practically unlimited relative motions between the molecules. The yielding behavior is due to the fact that, instead of crystals, structures consisting of eleven molecules arranged at the nodes of a space lattice with pentagonal sides are formed, referred as *quasi nuclei*, which don't fulfill totally the space and leave between them empty holes. Molecules acquiring a kinetic energy due to increased temperature migrate continuously from the quasi nuclei to the holes and new holes are formed at the empty positions left from molecules. The internal energy of the liquid equals the sum of the negative potential energy of the field of the attractive forces between the molecules and the positive kinetic energy of the migrating of them which is represented by the temperature. An activation internal energy density $o^{\varepsilon}$ corresponds to the melting temperature ${}_{o}T$ which defines the phase transition from the solid to the liquid state.

The deformability of the liquids is governed by the following two constitutive laws. The law of the compressibility

$${}_{V}\varepsilon_1 = \beta^{\bar{e}} \, {}_{a}\sigma^{e} \tag{2.54}$$

and the law of the shear viscosity

$$(\sigma_i{}^{\bar{J}})_{\bar{k}} = \left(\eta^{e}(\gamma_{i\bar{j}})^{\bar{k}}\right)_{\bar{k}}, \qquad (i \neq j) \tag{2.55}$$

where ${}_{V}\varepsilon_1$ is the volumetric strain, ${}_{a}\sigma^{e}$ the mean compressive normal stress (in the existing literature is represented by the pressure $p = -\,{}_{a}\sigma^{e}$), $\sigma_i{}^{j}$ the shear stresses, $\gamma_{i\bar{j}}$ the shear stain rates, $\beta^{\bar{e}}$ the compressibility coefficient and $\eta^{e}$ the shear viscosity. Eq. (2.54) is identical with Eq. (2.49) taking into account that the compressibility coefficient $\beta^{e}$ equals the inverse of the balk modulus $\bar{K}^{e}$. Both constants are related with the velocity of the sound in the liquid ${}_{l}c_i$ by the expression

$$K^{e} = \overline{\beta^{\bar{e}}} = {}_{l}C_i^2 \rho_1^{\bar{e}} \tag{2.56}$$

where $\rho_1^{\bar{e}}$ is the mass density.

The value of the shear viscosity $\eta^{e}$ depends on the temperature and is given by the relation

$$\eta^{e} = {}_{o}\eta^{e} exp\left(\frac{o\varepsilon}{kT}\right) \tag{2.57}$$

where

| | |
|---|---|
| ${}_{o}\varepsilon$ | is the activation energy of the molecules |
| $T$ | is the absolute temperature in ${}^{o}K$ |
| $k = 1.38 \times 10^{-23} J/K,$ | is the Boltzmann' s constant |
| ${}_{o}\eta^{e}$ | is the viscosity for ${}_{o}\varepsilon = 0$ |

From Eq. (2.57) results that the viscosity of the liquids decreases for increasing temperature.

Tensile stresses are not considered for the liquids except those developed in the tangential directions of the interfaces with the adjacent gaseous media, as for example the free surfaces and the surfaces surrounding drops and bubbles. The free surfaces are considered as membranes of zero thickness although the actual thickness equals that of the surface layer consisting of quasi – nuclei with six to seven, instead of eleven, molecules. Pure tensile stresses are developed in membranes having an axisymmetric or a cylindrical shape. For all other shapes shear stresses except the tensile ones are also developed implying that for the tensile behavior the complete set of stiffness constants $E^{e}, G^{e}$ and $v_1$ for isotropic materials contributes to the defined for liquids.

Anomalous attenuation observed at high frequencies ultrasonic measurements cannot be explained in terms of thermal conduction and shear viscosity alone and it is justified by the acception of a *bulk viscosity* or *viscous compressibility* ${}_{\eta}\beta^{\bar{e}}$ with values experimentally derived. Under the above assumptions Eqs. (2.54) and (2.55) take the following complete expressions

$$ {}_{a}\sigma^{e} = \overline{\beta^{\bar{e}}}\, {}_{V}\varepsilon_1 + {}_{\eta}\overline{\beta^{\bar{e}}}\, {}_{V}\gamma_1 \tag{2.58}$$

$$ \sigma_{i}^{\bar{j}} = G^{e}\varepsilon_{i\bar{j}} + \eta^{e}\gamma_{i\bar{j}} \tag{2.59}$$

## 2.8 Constitutive Laws of Gauges

For temperatures higher than that corresponding to the boiling point the atoms and molecules acquire high kinetic energies and the material change to the gaseous state. In gases the mean distance between molecules under standard conditions $\left( {}_0p^e = 1atm, \quad {}_0T = 0^oC = 273.16\ ^oK \right)$ is approximately twelve times higher than that occurring in the solid and the liquid states. For these distances the attractive forces between the particles are negligible and the internal energy of the gas equals the sum of the internal potential energy of the atoms or molecules plus their kinetic energy. Although the concentration $n$, representing the amount of molecules per unit volume, in gases is $12^3$~1600 times lower than that of the solid or the liquid state, is sufficiently huge in order the gases to be considered as homogeneous an disotropic deformable continuous media. The constitutive laws are those defined by Eqs. (2.58) and (2.59) for liquids, where the stiffness constants have the following expressions:

The bulk modulus or the compressibility coefficient for isothermal processes is

$$ {}_{is}K^e = {}_{is}\overline{\beta^{\bar{e}}} = p^e = nkTexp\left(-{}_{ex}U/kT\right) \tag{2.60a}$$

while for adiabatic processes is

$$ {}_{ad}K^e = {}_{ad}\overline{\beta^{\bar{e}}} = \gamma_1 p^e \tag{2.60b}$$

and the shear viscosity is

$$ \eta^e = \frac{1}{4\pi d^2}\sqrt{\frac{kT}{3}} \tag{2.61}$$

where

| | |
|---|---|
| $p^e$: | the pressure |
| $n^{\bar{e}} = p^{\bar{e}} /\ {}_0m$ : | the concentration of molecules |
| $\rho^{\bar{e}}$: | the mass denity |

| | |
|---|---|
| $k = 1.38 \times 10^{-23}\ (J/K)$: | the Boltzmann's constant |
| $T$ : | the absolute temperature $\left( {}^{0}K \right)$ |
| ${}_{ex}U$: | the potential of an external force field |
| $d$ : | the diameter of molecules |
| ${}_{0}m$: | the mass of molecules |
| $\gamma = {}_{p}c/\ {}_{v}c$ : | the specific heat ratio |
| ${}_{p}c$ : | the specific heat for isobaric processes ($p^e$= constant) |
| ${}_{v}c$ : | the specific heat for isochoric processes ($V^e$ = constant) |

For monatomic molecules where only their translational kinetic energy is contained in the internal energy of the gas, the specific heats are represented by the expressions

$$ {}_{p}c = \frac{5k}{2\ {}_{0}m}, \qquad {}_{v}c = \frac{3k}{2\ {}_{0}m}, \qquad \gamma = \frac{5}{3} \tag{2.62} $$

and are independent of the temperature. For polyatomic molecules, where in the internal energy the rotational kinetic energy, governed by the laws of quantum mechanics, is moreover contained, the specific heats depend on the temperature through extremely complicated expressions and their values are experimentally derived.

From Eq. (2.61) results that the dependence of the shear viscosity of the gases from temperature differs from that resulting for liquids as it increases for increasing temperature.

## 2.9 A Vacuum Hypothesis

The space not occupied by massive substances is usually considered as empty. Nevertheless to the empty space, or vacuum, a dielectric constant ${}_0\varepsilon$ is attributed implying the development of dipole electric moments under the action of electric fields, referred in Quantum Electrodynamics as Vacuum Polarization, which is accompanied by the experimentally observed spontaneous creation and annihilation of electron-positron pairs. This property permits the vacuum to be considered as an aggregate consisting of massless electrically neutral corpuscles in close packing at a spacing of the order of the Planck's length of $10^{-35}$ m which for the size of the elementary particles in the scale of $10^{-10}$ to $10^{-20}$m, can be considered as a homogeneous isotropic deformable continuum characterized by an energy density $\mu_{\bar{1}}^{\bar{e}}$ instead of the equivalent to it mass density $\rho_1{}^{\bar{e}} = -\mu_{\bar{1}}^{\bar{e}}/c^2$ , where $c$ is the speed of the light in the vacuum.

Electromagnetic waves, which are transversal (shear) waves, are propagated through the vacuum with the speed of the light $c$. Propagation of shear waves occurs exclusively in solid media and from that point of view the vacuum has to be considered as a peculiar solid medium consisting of energy, instead of mass, constituents. According to Eq. (2.53) the speed of propagation fulfills the relation

$$c^2 = \frac{G^e}{\rho_1{}^{\bar{e}}} = -\frac{c^2 G^e}{\mu_{\bar{1}}^{\bar{e}}} \qquad \text{or} \qquad G^e = -\mu_{\bar{1}}^{\bar{e}} \qquad (2.63)$$

From Eq. (2.63) results that the shear modulus $G^e$of the peculiar solid of the vacuum has the value of the energy density $\mu_{\bar{1}}^{\bar{e}}$.

The similarity between the expressions of the Newton's law of gravitation and Coulomb's law of the electric fields justifies the assumption that the gravitational constant , in correspondence with the dielectric constant ${}_0\varepsilon$ , has to be included to the properties of the vacuum. Under this view Eq. (2.63) represents likewise the speed of propagation of the transversal gravitational waves.

For this type of continuum the propagation of longitudinal (pressure) waves, like the sound ones, cannot be excluded. Although measurements of their speed of propagation ${}_{l}c$ are not yet available, this is assumed to be equal to that of the speed of light, because this is considered as the maximum possible speed in the nature. Taking into consideration the equation (2.53), the assumption

$$ {}_{l}c^2 = \frac{C^e}{\rho_1^{\bar{e}}} = \frac{c^2 C^e}{\mu_1^{\bar{e}}} = c^2 = \frac{c^2 G^e}{\mu_1^{\bar{e}}} \tag{2.64} $$

implies, according to the Eqs. (2.47a, b).

$$ C^e = \frac{(1-\nu_1)E^e}{(1+\nu_1)(1-2\nu_1)} = G^e = \frac{E^e}{2(1+\nu_1)} = \mu_1^{\bar{e}} \tag{2.64a} $$

which is impossible for any value of $0 \leq \nu_1 \leq 0.5$ and thus contradicts to the hypothesis of the peculiar solid for the vacuum, which constitutes a necessary condition for the propagation of the transversal electromagnetic waves.

For $0 \leq \nu_1 \leq 0.5$ results that $C^e \geq 2G^e$ and ${}_{l}c \geq \sqrt{2}c$. The maximum condensation of the energy corpuscles constituting the microstructure of the peculiar solid results for their arrangement in a tetrahedral space lattice with minimum possible distance between the nodes equal to the Planck's length $l_p = 1.616 \times 10^{-35} m$.

Two alternative considerations for the tissue of the lattice of the energy corpuscles forming the microstructure of the vacuum are possible. The one considering the corpuscles as spheres in close contact. And the other which considers them as connecting elements or strings (see Chapter 4) arranged along the edges of the tetrahedra and converging at the geometrical points of the nodes of the lattice, similar to the bars of a space truss macrostructure. In the latter case the zero dimensional nodes are considered as solid bodies fulfilling the equilibrium condition between the reactions developed at the end points of the connecting elements converging to the node as well as the compatibility condition between their end displacements. In both cases the propagation of electromagnetic and gravitational waves through the elastic lattice of the corpuscles is possible. The mutual emission and absorption of energy between interacting elementary particles is executed through their contact with the distinct nodes of the lattice. The effect of this procedure is equivalent with that consisting to the exchange of photons and gravitons moving in empty space between interacting particles. (Fig. 2.4)

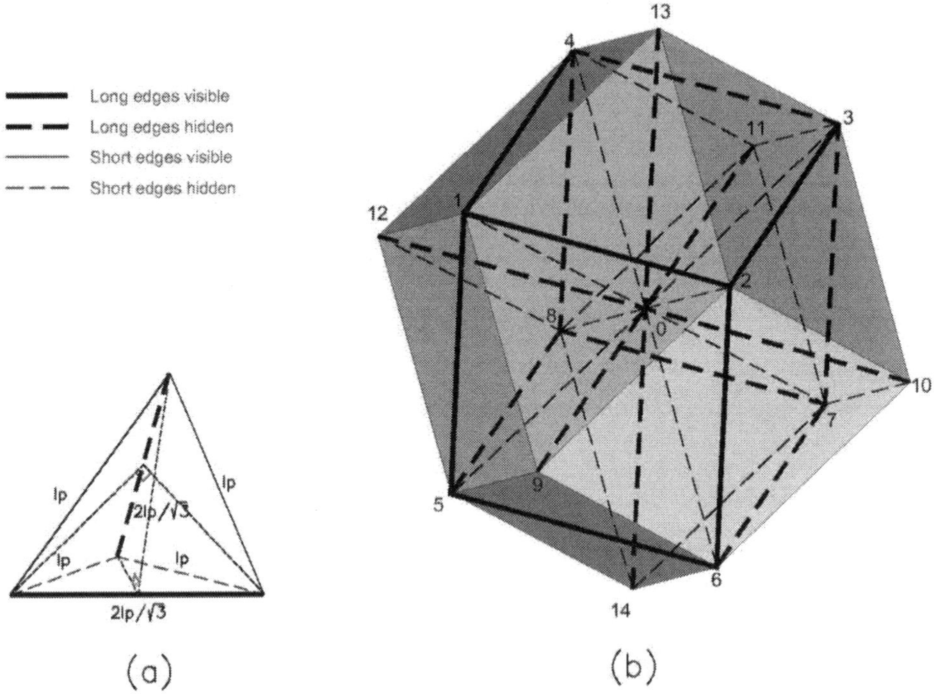


*Figure 2.4 - Orthotetrahedron (a) and Rombic Dodecahedron (b)*

The space around a node of the lattice is completely fulfilled by a set of 24 *orthotetrahedra* (Figs. 2.4a, b) having right dihedral angles at two opposite edges of length $(2\sqrt{3})l_p = 1.1547\ l_p$, the length of the rest four edges equal to $l_p$, the rigid angles at the summits equal to $\pi/6$ and the volume equal to $2\sqrt{3}(l_p/3)^3 = 0.1283\ l_p^3$. For the corresponding space lattice considered in a macroscopic scale as an isotropic deformable solid continuum a value $\nu_1 = 1/3$ of the Poisson's ration results, after applying a structural mechanics approach.

For that case the elastic constants receive the values

$$E^e = \frac{8}{3}\mu_{\bar{1}}^{\bar{e}}, \qquad G^e = \mu_{\bar{1}}^{\bar{e}} = \frac{3}{8}E^e, \qquad C^e = \frac{3}{2}E^e = 4\mu_{\bar{1}}^{\bar{e}} \qquad (2.65)$$

and

$$_{l}c = 2c \tag{2.66}$$

The propagation of longitudinal electromagnetic or gravitational, in the case they actually exist, is expected to be extremely difficult to be detected because of their small amplitudes and their higher frequencies.

From the value of the cosmological constant $\Lambda = 8\pi(g/c^4)\mu^{\bar{e}} = 3{,}1 \times 10^{-53} m^{-2} s^4$, (where $g = 6.6726 \times 10^{-11}\ kg^{-1} m^3 s^2$ the gravitational constant) which is estimated for the accelerated expansion of the universe recently observed, a value of the energy density $\mu_{\bar{1}}^{\bar{e}} = 1.5 \times 10^{-10} kg\ m^{-1} s^2 (1.47 \times 10^{-9} Nm^{-2})$ results corresponding to a mass density $\rho_{\bar{1}}^{\bar{e}} = 1.67 \times 10^{-27}\ kg\ m^{-3}$. This extremely low value of the energy density implies that the peculiar solid representing the vacuum is an "ultra soft" and "ultra flexible" substance.

# 3. THE PARTICLE

## 3.1 The Concept of Particle

As particle is considered a non-infinitesimal volume segment of a continuum in the case the motion of its constituent material points are functions of their position in an orthogonal reference system multiplied by single motion parameters of the whole particle referred likewise to that reference system. In that case these overall motion parameters are related with the resultants of the forces and momentums applied on the material points, which are referred likewise in the same reference system. Basic requirement for this replacement of the system of the distributed forces and momentums and the conjugate to them displacements and velocities by single overall state parameters is the following condition: The work done by the totality of the distributed forces or the kinetic energy developed by the totality of the distributed momentums for the corresponding displacements or velocities respectively to be equal to the work or the kinetic energy of their resultants for the corresponding overall motion parameters.

This concept of the particle attributed to deformable continuum segments associated with an overall motion or deformation mode of the material points contained which are subjected momentums or forces applied on them concerns a wide variety of physical systems seemingly of entirely different nature. The following kinds of such systems can be considered.

a. Deformable continuum segments of a finite volume surrounded by a closed surface. In the case of solid material any set of forces and momentums applied can be considered associated with the corresponding to them deformation and motion mode. For the case of particles of fluid material, as they are for example the drops, the pebbles etc., their form is due and closely associated with a specific set of external pressures and internal stresses.

A rigid un-deformable body where the motion of its material points is completely determined by the translation of its center of mass and the rotations around the axes of the adapted to it orthogonal reference system, is the most common example.

b. Continuous systems of liquid material points executing trajectories with finite orbital velocities having values tending asymptotically to zero at infinite distances from the point appearing the maximum kinetic energy density, which considered as the center of the system. Although to those systems finite values of the total angular momentum and of the total kinetic energy are attributed, they occupy a theoretically infinite volume and they are created and recognized exclusively from the mode of the overall motion. Characteristic examples are the vortex filaments, the vortex rings, the spherical vortices etc.

As a limiting case is referred the elementary particles of quantum physics with their state determined by a finite set of quantum numbers.

The target of the present chapter is restricted to the investigation of the nature of the overall motion parameters and the conjugate with them resultants of the forces and momentums applied on the totality of its material points. This investigation becomes simpler in the case of continuum considered in the laminar state where the forces applied and the motions performed by the material points are represented by polar vectors $f_p$ and $u_p$ respectively. From the octal decomposition presented in the paragraph 3.5 results that the overall parameters determining the state of the particle are elements of octonions, which contain polar and axial vectors and moreover the innovative concepts of real and imaginary scalar forces and displacements.

### 3.2 Motives, Composants, Shape and Lever-arm Functions

The displacements of the material points, represented by the polar vectors

$$u = \sum_i u_i$$

can be expressed as functions $\zeta_{\bar{I}i}[\subset r_i]$ of the components $r_i$ of the position vector

$$r = \sum_i r_i$$

multiplied by a single octonion parameter

$$U = \sum_I U_I,$$

which describes their total motion, according to the relation

$$u = \sum_i u_i = \sum_i \sum_I \left( U_I \zeta_{\bar{I}i} \left[ \subset_i r_i \right] \right)_i = \sum_I \left( U_I \sum_i \zeta_{\bar{I}i} \right) \qquad (3.1)$$

Where $i = x, y, z$ and $I$, anyone of the eight indices of the octonion components.

For example, for a rigid body translation $U_i$ of the particle, the translations of the material particles are $u_i = U_i$, $\zeta_{\bar{\iota}i} = 1$, and $u_j = u_k = 0$, $\zeta_{\bar{\iota}j} = \zeta_{\bar{\iota}k} = 0. u_j = u_k = 0$, $\zeta_{\bar{i}j} = \zeta_{\bar{i}k} = 0$. For a rigid body rotation $U^i$ of the particle around the $e^i$ axis, the displacements of the particles are

$$u = U^i \otimes r = \left(U^i r_j\right)_k + \left(U^i r_k\right)_{\bar{j}} = u_k + u_j$$

and

$$\left(\zeta^{\bar{\iota}}{}_i\right)_{\bar{e}} = 0, \quad \left(\zeta^{\bar{\iota}}{}_k\right)_j = r_j, \quad \left(\zeta^{\bar{\iota}}{}_j\right)_{\bar{k}} = r_k$$

The single octonion parameter $U$ of the total motion of the particle will be referred in the sequence with the term *motion*, the octonion components $U_I$ of $U$ with the term *motives* and the functions

$$\zeta_{\bar{I}i} \left[ \subset_i r_i \right]$$

with the term *shape functions*.

On the other hand, the system of the local forces

$$_n f = \sum_i {}_n f_i$$

applied on the material points $n$ of the element, can be considered as components of a total resultant, represented by the octonion

$$F = \sum_I F_I,$$

according to the relation

$$F = \sum_I F_I = \sum_I \sum_i (\xi_{I\bar{\imath}} f_i) \tag{3.2}$$

where

$$\xi_{I\bar{\imath}} = \xi_{I\bar{\imath}}\left[\underset{i}{\subset} r_i\right]$$

are functions of the position vector $r$, which will be referred as *lever-arm functions*. For example the polar vector resultants are

$$F_i = \sum_n {}_n f_i$$

and

$$\xi_{i\bar{\imath}} = 1, \quad \xi_{i\bar{\jmath}} = \xi_{i\bar{k}} = 0.$$

The axial vector resultant or moment $F^i$ with respect to the $e^i$ axis is

$$F^i = \sum_n \left( \left( {}_n r_j \, {}_n f_k \right)^i + \left( {}_n r_k \, {}_n f_j \right)^{\bar{\imath}} \right)$$

and

$$\left(\xi^i{}_{\bar{\imath}}\right)_{\bar{e}} = 0, \quad \left(\xi^i{}_{\bar{\jmath}}\right)_{\bar{k}} = r_k, \quad \left(\xi^i{}_{\bar{k}}\right)_j = r_j$$

In order to be avoided any confusion with the forces $f$ applied on the material points, which are referred as local components of the resultant $F$, the octonion components $F_I$ of $F$ will be referred as *composants*.

The shape and lever-arm functions $\zeta_{\bar{I}i}$ and $\xi_{I\bar{\imath}}$ are elements of two 8X3 dimensional tensors $\zeta$ and $\xi$ respectively.

For the deformable continuum was assumed that the local forces and displacements belong to irrotational polar vector fields which, for conservative systems, have as state variable a scalar potential, whose the variation equals the work. Consequently, criterion of the complete equivalence between the system of the local forces and displacements and its representation by the corresponding composants and motives, consists to the equality of the work resulting for the two representations, which, according to the Eq. (3.1), is expressed by the equation

$$W = -\sum_i u_i f_i = -\sum_i \sum_I (U_I \zeta_{\bar{I}i})_i f_i = \\ = -\sum_I U_I \sum_i (\zeta_{\bar{I}i} f_i)_I = -\sum_I U_I F_I \tag{3.3}$$

From (3.3) results

$$F_I = \sum_i (\zeta_{\bar{I}i} f_i) \tag{3.4}$$

and, from the comparison with Eq. (3.2),

$$\zeta_{\bar{I}i} = \xi_{I\bar{\imath}} \tag{3.5}$$

Eq. (3.5) is the analytical expression of the following fundamental theorem relating composants with motives.

*Theorem 1. The lever-arm functions $\xi_{I\bar{\imath}}$of the composants $F_I$ are identical with the shape functions $\zeta_{\bar{I}i}$ of the conjugate motive $U_I$.*

The above fundamental theorem is of great importance, because it affects directly the definitions of the various composants, as they have to be related with the mode of the motion of the system, containing and its deformation, which is determined by the shape functions of the conjugate motive. For example, the overall accepted definition of the moment with the simple linear expression of the cross product $r \times f$, is directly related with rotations of rigid bodies, for which the displacements of the material points appear a similar linear distribution. But for non-rigid, deformable bodies, a total rotation does not imply necessarily a linear distribution of the displacements and, accordingly, in that case the moments have to be defined by a more complicated non-linear expression. This will become apparent in the next sections, together with the necessity for the acceptance of real and imaginary *scalar* composants and motives, beyond the self-evident existence of those having the form of polar and axial vectors.

For the kinematic case the parameters corresponding to

$$u = \sum_i u_i, \quad f = \sum_i f_i, \quad U = \sum_I U_I, \quad F = \sum_I F_I,$$

are respectively,

the local velocities $v = \sum_i v_i,$

the local momentums $p = \sum_i p_i,$

the system velocity $V = \sum_I V_I$

and the momentum resultant $P = \sum_I P_I$

They fulfill the following equations, similar to the Eqs. (3.1) and (3.2)

$$v = \sum_i v_i = \sum_i \sum_I (V_I \zeta_{\bar{I}i}) \quad (3.1a)$$

$$P = \sum_I P_I = \sum_I \sum_i (\xi_{I\bar{i}} P_i) \quad (3.2a)$$

as well as the following expression for the kinetic energy, similar to the Eq. (3.3)

$$T = -\frac{1}{2}\sum_i v_i p_i = -\frac{1}{2}\sum\sum (V_I \zeta_{\bar{I}i})_i p_i =$$
$$= -\frac{1}{2}\sum_I V_I \sum_i (\zeta_{\bar{I}i} p_i)_I = -\frac{1}{2}\sum_I V_I P_I \qquad (3.3a)$$

The shape and lever-arm functions for the kinematic case are identical to those of the static case.

## 3.3 Invariants

The modules $|f_p|$ and $|u_p|$ of the polar vectors representing the forces applied and the displacements developed respectively on the material points of the continuum considered in the laminar state are invariant under orthogonal transformations. The motion $U$ and the resultant $F$ of the particle are octonions which are analysed in scalar and vector components of the form

$$U = U_1 + U_P + U^A + U^e \qquad (3.6a)$$
$$F = F_1 + F_P + F^A + F^e \qquad (3.6b)$$

The scalar entities $U_1, |U_P|, |U^A|, U^e$ and $F_1, |F_P|, |F^A|, F^e$ have also to be invariant under orthogonal transformations. According to the relation (3.2) the scalar and vector components of the resultant $F$ have the form

$$F_1 = \sum_i ((\xi_{1\bar{\imath}})_{\bar{\imath}} f_i)^1 = \left((\xi_{1\bar{p}})_{\hat{p}} f_p\right)^1 \qquad (3.7a)$$

$$F_P = \sum_j F_{\bar{j}} = \sum_j \sum_i \left((\xi_{j\bar{\imath}}) f_i\right)_j = \sum_i ((\xi_{P\bar{\imath}}) f_i)_P = (\xi_{P\hat{p}} f_p)_P \qquad (3.7\,b)$$

$$F^A = \sum_j F^j = \sum_j \sum_i \left((\xi^j{}_{\bar{\imath}}) f_i\right)^j = \sum_i \left((\xi^A{}_{\bar{\imath}}) f_i\right)^A = \left(\xi^A{}_{\hat{p}} f_p\right)^A \qquad (3.7c)$$

$$F^e = \sum_i \left((\xi^e{}_{\bar{\imath}})^i f_i\right)^e = \left((\xi^e{}_{\hat{p}})^a f_p\right)^e \qquad (3.7d)$$

where

$$\xi_{P\hat{\imath}} = \sum_j \xi_{j\bar{\imath}}, \qquad \xi^A{}_{\bar{\imath}} = \sum_j \xi^j{}_{\bar{\imath}},$$

and

$$\xi_{1\hat{p}} = \sum_i \xi_{1\bar{\imath}} = \zeta_{\bar{1}p}, \qquad \zeta_{P\hat{p}} = \sum_i \xi_{P\bar{\imath}} = \zeta_{\hat{P}p},$$

$$\xi^A{}_{\hat{p}} = \sum_i \xi^A{}_{\bar{\imath}} = \zeta^{\hat{A}}{}_p, \qquad \xi^e{}_{\hat{p}} = \sum_i \xi^e{}_{\hat{\imath}} = \zeta^{\bar{e}}{}_p$$

From Eqs. (3.7) result

$$\xi^1{}_{\hat{p}} = F^1 \hat{f}_p, \qquad \xi_{P\hat{p}} = F_P \hat{f}_p = F_P \hat{f}_p = \left(F_P \cdot \hat{f}_p\right)^1 + \left(F_P \times \hat{f}_p\right),$$

$$\xi^A{}_{\hat{p}} = F^A \hat{f}_p = \left(F^A \circ \hat{f}_p\right)_e + \left(F^A \otimes \hat{f}_p\right), \qquad \xi_{e\hat{p}} = F_e \hat{f}_p \tag{3.8}$$

and

$$\left(\xi_{1\hat{p}}\right)^2 = F_1{}^2 \hat{f}_p^2, \qquad \left(\xi_{P\hat{p}}\right)^2 = \left(F_P \cdot \hat{f}_p\right)^2 + \left(F_P \times \hat{f}_p\right)^2,$$

$$\left(\xi^A{}_{\hat{p}}\right)^2 = \left(F^A \circ \hat{f}_p\right)^2 + \left(F^A \otimes \hat{f}_p\right)^2, \qquad \left(\xi^e{}_{\hat{p}}\right)^2 = F^{e2} \hat{f}_p^2 \tag{3.9}$$

The terms at the right side of the relations (3.9) are invariant under orthogonal transformations, this implying the invariance of the expressions $\left(\xi_{R\hat{p}}\right)^2 = \left(\zeta_{\hat{R}p}\right)^2$ where by $R$ is represented any one of the octonion indices $\left({}_1, P, {}^A, {}^e\right)$. The expressions $\xi_{I\bar{\imath}} = \zeta_{\bar{I}i}$ are functions of the components $x, y, z$ of the position vector $r$, which are not invariant under orthogonal transformations, while the module $|r|$ or $r$ is invariant. From the relations (3.9) results the following *invariance condition:*

*Theorem 2. The expressions* $\xi_{1\hat{p}} = \zeta_{\bar{1}p}$, $\xi_{P\hat{p}} = \zeta_{\hat{P}p}$, $\xi^A{}_{\hat{p}} = \zeta^{\hat{A}}{}_p$ *and* $\xi^e{}_{\hat{p}} = \zeta^{\bar{e}}{}_p$ *are functions of the position vector* $r$ *and not of its components* $x, y, z$.

The polynomial expansion of these functions have the form

$$\xi_{R\hat{p}} = \zeta_{\hat{R}p} = {}_0\beta + {}_1\beta r^2 + {}_2\beta r^2 + \cdots + {}_{2i}\beta r^{2i} + {}_{2i+1}\beta r^{2i+1} + \cdots =$$

$$= \left( {}_0\beta + \sum_i {}_{2i}\beta r^{2i} \right)_1 + \left( {}_1\beta + \sum_i {}_{2i+1}\beta^{2i} \right)_1 r \qquad (3.10)$$

where ${}_i\beta$ are scalar coefficients and

$$\xi^2_{R\hat{p}} = \zeta^2_{\hat{R}p} = \left( {}_0\beta + \sum_i {}_{2i}\beta r^{2i} \right)^2 + \left( {}_1\beta + \sum_i {}_{2i+1}\beta r^{2i} \right)^2 r^2 \qquad (3.11)$$

which is invariant under orthogonal transformations. The lever-arm and shape functions constituting the term with the lowest degree of $r$ in the polynomial expansion are considered as the *fundamental* functions.

The target of the present treatise is restricted to the investigation of the fundamental lever-arm and shape functions which are sufficient for the purpose the nature of the corresponding resultants and motives to become apparent. The terms of higher degree are determined through appropriate procedures for each one specific case, which requires the introduction of the constitutive law of the materials and the application of the general principles of Physics. An example of such a procedure in contained in Chapter 4 for the case of the one dimensional aggregate sequence of particles.

### 3.4 Conformal and Contra-formal Motions

A three dimensional network of straight lines parallel to the unit vectors $e_l, e_m, e_n$ of an orthogonal basis, which satisfy the relations

$$e_l \cdot e_m = e_m \cdot e_n = e_n \cdot e_l = 0 \qquad (3.12)$$

is attached to the finite continuum segment in the state of tranquility.

The vector equations of the three straight lines of the network passing through the point ${}_0r$ and parallel to the unit vectors $e_l, e_m$ and $e_n$ are respectively

$$ {}_{l}r = {}_{0}r + \lambda_1 e_l, \qquad {}_{m}r = {}_{0}r + \mu_1 e_m, \qquad {}_{n}r = {}_{0}r + \nu_1 e_n \tag{3.13} $$

where $\lambda_1, \mu_1, \nu_1$ are real scalar parameters.

After a motion of the continuum, defined by the displacement function $u[r]$, the orthogonal network of straight lines is transformed to a network of curves, represented by the equations (Fig 3.1)

$$ {}_{l}r' = {}_{l}r + u[{}_{l}r] = {}_{0}r + \lambda_1 e_l + u[{}_{l}r] $$

$$ {}_{m}r = {}_{m}r + u[{}_{m}r] = {}_{0}r + \mu_1 e_m + u[{}_{m}r] \tag{3.14} $$

$$ {}_{n}r = {}_{n}r + u[{}_{n}r] = {}_{0}r + \mu_1 e_n + u[{}_{m}r] $$

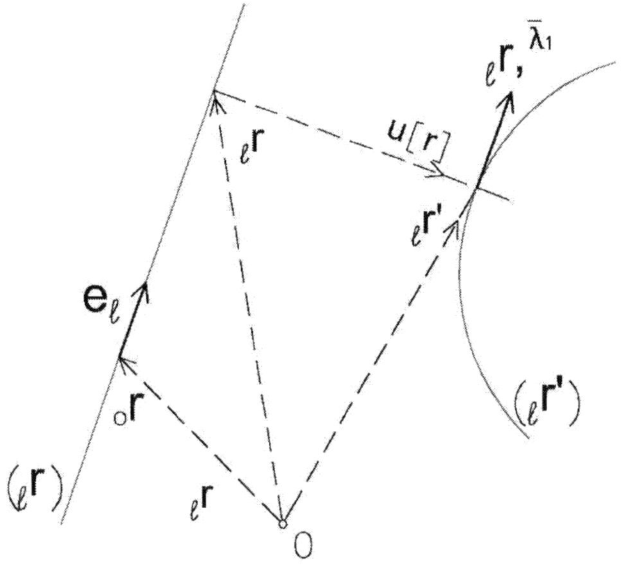


*Figure 3.1 – Deformation of straight lines to curves*

The tangent vectors of the curves at the point $r'$ are

$${}_{l}r',^{\lambda_1} \; ,= e_l + u,^{\lambda_1} \; = e_l + u_{,\bar{l}} \; {}_{l}r,^{\lambda_1} \; = e_l + u_{,\bar{l}} \, e_l$$

$${}_{m}r',^{\mu_1} \; ,= e_m + u,^{\mu_1} \; = e_m + u_{,\bar{m}} {}_{m}r,^{\mu_1} \; = e_m + u_{,\bar{m}} \, e_m$$

$${}_{n}r',^{\nu_1} \; ,= e_n + u,^{\nu_1} \; = e_n + u_{,\bar{n}} \; {}_{n}r,^{\nu_1} \; = e_n + u_{,\bar{n}} \, e_n \qquad (3.15)$$

The orthogonality condition for the tangent vectors ${}_{l}r',^{\lambda_1}$ and ${}_{m}r',^{\mu_1}$ at the deformed lines ${}_{l}r'$ and ${}_{m}r'$ respectively of the network, after taking into account Eqs. (3.12), is

$${}_{l}r'^{\lambda_1} \cdot {}_{m}r',^{\lambda_1} = (u_{,\bar{l}} \, e_l) \cdot e_m + e_l \cdot (u_{,\bar{m}} \, e_m) + (u_{,\bar{l}} \, e_l) \cdot (u_{,\bar{m}} \, e_m) = 0 \qquad (3.16)$$

For infinitesimal displacements the last term at the right part is omitted as second order infinitesimal and the orthogonality condition has the form

$$(u_{,\bar{l}} \; e_l) \cdot e_m + e_l \cdot (u_{,\bar{m}} \; e_m) = 0 \qquad (3.16a)$$

The motion specified by displacement functions $u[r]$which satisfy the conditions (3.16) or (3.16a), furthermore referred as *conformity condition*, the three pairs of unit vectors of any orthogonal basis throughout all points of the continuum, are characterized as *conformal*, denoted by the symbol ${}_{c}u$ through a low left index $c$. Otherwise they are characterized as *contra-formal* and are denoted by the symbol ${}^{c}u$ through an upper left index $c$. As conformal and contra-formal are characterized and the relevant to the function $u$ motives $U_I$, the composants $F_I$ and the corresponding to them shape and lever-arm functions and they are likewise denoted by the same low and upper left indices $c$ .
It is evident that the rigid body translations and rotations of a particle constitute fundamental conformal motions. Moreover displacement functions $u[r]$, with their derivatives $u_{,\bar{l}}$, $u_{,\bar{m}}$ , $u_{,\bar{n}}$ equal to a real scalar, are likewise covariant, because in that case Eq. (3.16a), taking into consideration and the Eqs. (3.12), takes the value,

$$\left(u_{,\bar{l}}\right)_1 (e_l \cdot e_m) + (e_l \cdot e_m)\left(u_{,\bar{m}}\right) = 0$$

Eq. (3.16a), multiplied by $(r_l\bar{e}_l)_1\ (r_m\bar{e}_m)_1$ and taking into account that $(r_i\bar{e}_i)_1\ e_i = r_i$ takes the form

$$(u_{,\bar{l}}\, r_{\bar{l}}) \cdot r_m + r_i \cdot (u_{,\bar{m}}\ r_m) = 0 \tag{3.16b}$$

For $u = u_l + u_m + u_n$ Eq. (2.3.19a) takes the analytical expression

$$\left(\left((u_{l,\bar{l}})e_l\right)_l + \left((u_{m,\bar{l}})^n\, e_l\right)_m + \left((u_{n,\bar{e}})^{\bar{m}}\, e_l\right)_n\right) \cdot e_m +$$
$$\left(\left((u_{l,\bar{m}})^{\bar{n}} e_m\right)_l + \left((u_{m,\bar{m}})_1\, e_m\right)_m + \left((u_{n,\bar{m}})^l e_m\right)_n\right) \cdot e_m = 0$$

or after canceling the zero dot products,

$$\left((u_{m,\bar{e}})^n\, e_l\right)_m \cdot e_m + \left((u_{l,\bar{m}})^{\bar{n}} e_m\right)_l \cdot e_l = (u_{m,\bar{l}} - u_{l,\bar{m}})(e_l e_m)^n = 0$$

and, finally, the conformity condition takes the form

$$u_{m,\bar{l}} = u_{l,\bar{m}} \tag{3.16c}$$

if $e_i, e_j, e_k$ is another orthogonal system, resulting from the system $e_l, e_m, e_n$ through an orthogonal transformation, will be

$$u = \sum_i u_i, \qquad u_i = \sum_l u_l e_{l\bar{i}}, \qquad r_m = \sum_j r_j e_{j\bar{m}}, \qquad r_{m,\bar{j}} = e_{j\bar{m}}$$

and

$$u_{i,\bar{j}} = \sum_l u_{l,\bar{j}} e_{l\bar{i}} = \sum_l \sum_m u_{l,\bar{m}} r_{m,\bar{j}} e_{l\bar{i}} = \sum_l \sum_m u_{l,\bar{m}} e_{l\bar{m}} e_{l\bar{i}}$$

$$u_{i,\bar{j}} = \sum_m u_{m,\bar{i}} e_{m\bar{j}} = \sum_m \sum_l u_{m,\bar{l}} r_{l,\bar{i}} e_{m\bar{j}} = \sum_m \sum_l u_{m,\bar{l}} e_{i\bar{l}} e_{m\bar{j}}$$

From these relations, after taking into consideration Eq. (3.16c) and the relation $e_{j\bar{m}} e_{l\bar{i}} = e_{i\bar{l}} e_{m\bar{j}}$, results

$$u_{i,\bar{j}} = u_{j,\bar{i}}$$

which constitutes the proof of the following theorem:

*Theorem 3: A displacement function* $u$ *which is conformal in a random orthogonal system, is likewise conformal in all orthogonal systems.*

### 3.5 The Octal Decomposition

The determination of the different kinds of composants and motives, as well as of the corresponding lever-arm and shape functions, with respect to a reference system *oxyz,* is simplified by considering the effect of a single polar vector force

$$f = \sum_i f_i = f_x + f_y + f_z$$

applied at a point of the particle.

$$_1r = \sum_i {_1r_i} = {_1x} + {_1y} + {_1z}$$

According to the following procedure becomes possible to replace the single force applied to the single point by the equivalent superposition of eight systems each one containing eight forces applied to eight different points in a manner for each one system to result non zero value for one composant $F_I$ of the resultant $F$ and zero values for the rest composants.

As points of application of forces are taken the eight points $_nr = \pm\, {_1x} \pm\, {_1y} \pm\, {_1z}$, ($n$=1 to 8), resulting for all possible combinations of the signs of the coordinates $\pm\, {_1r_i}$, which are the mirror points of $_1r$ with respect to the coordinate planes $e^x, e^y$ and $e^z$. At the seven points $_nr$, ($n$=2 to 8) are applied forces $_nf$ equal to the mirror vectors of $\frac{1}{8}f$ with respect to the coordinate planes, with components $_nf = \pm\frac{1}{8}f_x \pm \frac{1}{8}f_y + \frac{1}{8}f_z$ belonging to eight different combinations of the signs throughout the seven points, $_2r$ to $_8r$, while at the point $_1r$, for each one of the above combinations, is applied the force $_1f = \frac{1}{8}f$. The resulting eight systems of forces are presented in (Figs. 3.2a to h). The superposition of the eight systems results to a force $f$ applied at the point $_1r$ and zero forces at the rest points $_nr$, ($n$=2 to 8). This decomposition of the force $f$ to the eight systems of forces is unique and will be refereed as *octal decomposition.*

From the (Fig. 3.2) becomes apparent that to each one of the systems resulting from the octal decomposition corresponds exclusively a single composant. To the systems presented in the (Figs. 3.2b, c, d) correspond the force composants $F_x, F_y$ and $F_z$ respectively, while to the systems presented in the (Figs. 2.3.2. e, f, g) correspond the moment composants. $F^x, F^y$ and $F^z$ respectively.

The two systems of forces presented in the (Figs. 3.2a and h) correspond to zero polar and axial vector composants and for rigid body motions, translations or rotations, the total work of the local forces for the corresponding local displacements equals zero. This condition is satisfied in the case of local forces applied on a rigid un-deformable continuum. But in the general case of the deformable continuum there are modes of overall motion, due to deformation, implying local displacements of the points $_nr$ having directions equal to those of the local forces $_nf$. And the total work of the local forces for the local displacements must be equal to the work of a composant for the corresponding motive. These two pairs of composants and motives, which are not represented by polar or axial vectors, can be represented by scalar octonion components, a real one for the system of the (Fig. 3.2a) and an imaginary for the system of the (Fig. 3.2h), as this will be shown in the subsequent corresponding sections.

In the following sections, where the shape, lever-arm and displacement functions for all kinds of composants $F_I$ and motives $U_I$ are discussed in detail, will become apparent that the distinction in conformal and contraformal entities, is absolutely necessary for the complete description of the behavior of particles subjected a system of forces. This complete description of the behavior in terms of composants and motives requires finally at least sixteen parameters for each one kind of the above entities, eight conformal and eight contra-formal, each one set consisting of octonion components.

In (Figs. 3.2a1, b1, e1 and h1) are represented super-positions of the system of forces corresponding to the composants, $F_I$ which result to the total composants

$$F_P \sum_i F_i \quad \text{and} \quad F^A \sum_i F^i$$

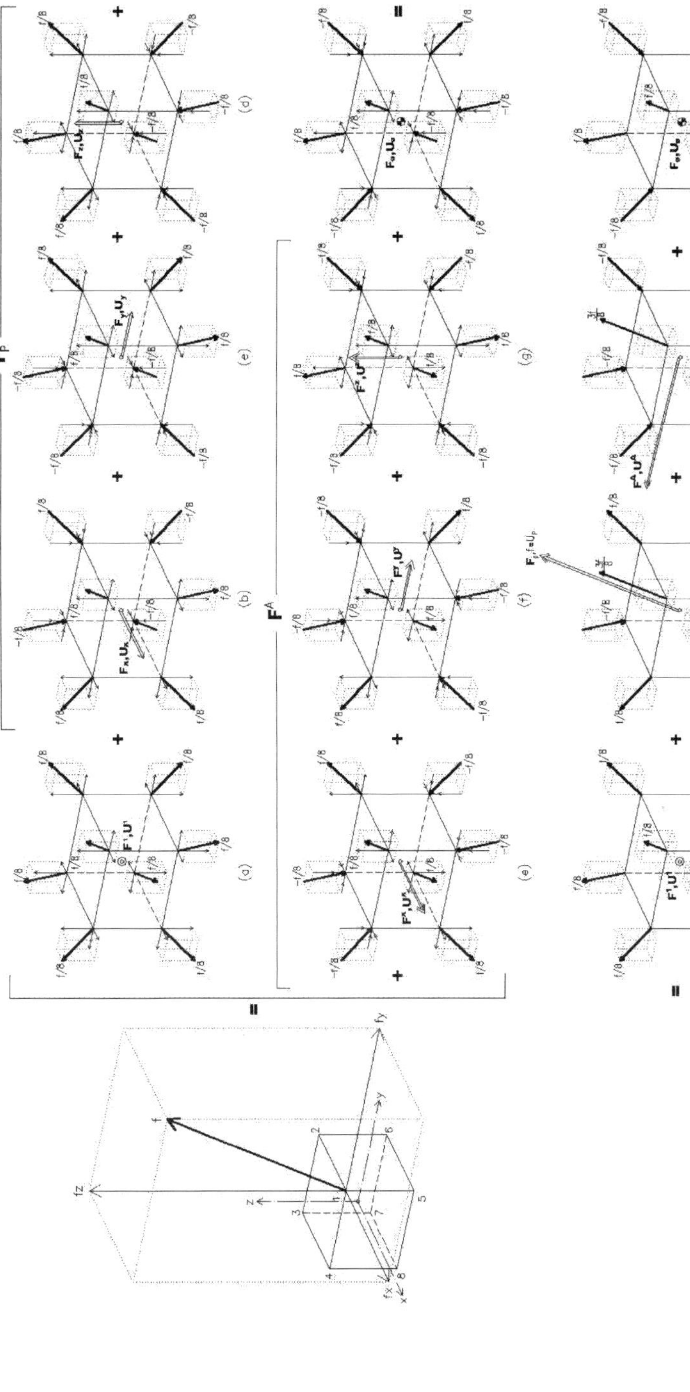


*Figure 3.2 - The octal Decomposition*

### 3.6 Polar Vector Composants and Motives

For the system of forces shown in (Fig. 3.2b) the vector sum of the forces ${}_n f_i$ equals the force composant in the direction of $e_x$ axis

$$F_x = \sum_{n=1}^{8} {}_n f_x = f_x, \qquad \sum_{n=1}^{8} {}_n f_y = \sum_{n=1}^{8} {}_n f_z = 0,$$

implying the fundamental functions

$$\xi_{x\bar{x}} = 1, \qquad \xi_{x\bar{y}} = \xi_{x\bar{z}} = 0$$

For a rigid body and hence a conformal translation in the same direction, defined by the motive $U_x$ , the local displacements of the points ${}_n r$ are

$${}_n u_x = {}_c U_x, \qquad {}_n u_y = {}_n u_z = 0, \qquad {}_c \zeta_{\bar{x}x} = 1, \qquad {}_c \zeta_{\bar{x}y} = {}_c \zeta_{\bar{x}z} = 0$$

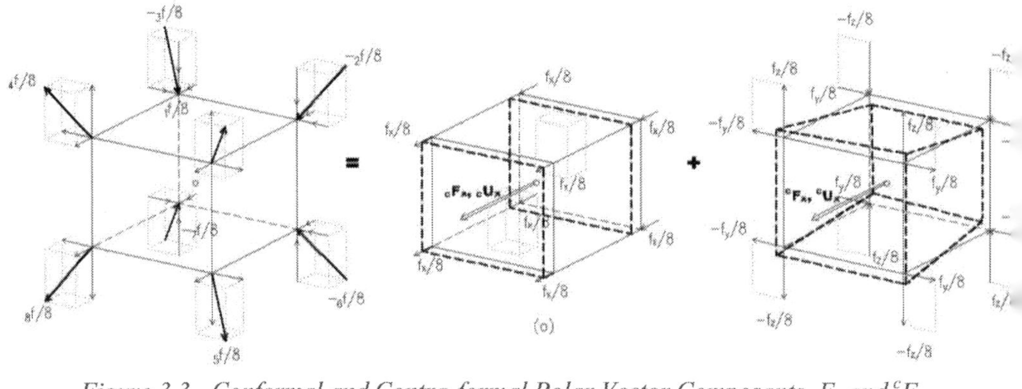

*Figure 3.3 - Conformal and Contra-formal Polar Vector Composants ${}_c F_x$ and ${}^c F_x$*

By similar approach result the relations for the systems of forces shown in (Figs. 3.2c and d), which for rigid body translations along the axes $y$ and $z$, correspond the composants and motives ${}_c F_y$, ${}_c U_y$ and ${}_c F_z$ , ${}_c U_z$ respectively, expressed by the following general relations:

$${}_c \zeta_{\bar{\imath}j} = {}_c \xi_{i\bar{\jmath}} = \delta_{ij} r_i \bar{r}_j, \quad \{ {}_c \zeta_{\bar{\imath}i} = {}_c \xi_{i\bar{\imath}} = 1, \quad {}_c \zeta_{\bar{\imath}j} = {}_c \xi_{i\bar{\jmath}} = 0 \;\; i \neq j \} \tag{3.18a}$$

$$ {}_cF_i = \sum_j \sum_{n=1}^{8} {}_{nci}\xi_{i\bar{j}}\, {}_nf_j = \sum_j \sum_{n=1}^{8} \left(\delta_{ij}\left({}_nr_i\, {}_n\bar{r}_j\right)^{\bar{k}}\, {}_nf_j\right) = \sum_{n=1}^{8} {}_nf_i = f_i z $$

(3.18b)

$$ {}_iu_i = {}_cU_i\ {}_{ci}\zeta_{\bar{\imath}i} = {}_cU_i\ , \qquad u_j = {}_cU_i\ {}_c\zeta_{\bar{\imath}\bar{j}} = 0\ \{i \neq j\} \qquad (3.18c) $$

$$ {}_cF_p = \sum {}_cF_i $$

The motion ${}_cU_P$ and the total resultant ${}_cF_P$ for a rigid body translation are represented by the polar vectors

$$ {}_cU_P = \sum_i {}_cU_i, \qquad {}_cF_P = \sum_i {}_cF_i = \sum_i f_i = f \qquad (3.18d) $$

From Eqs. (3.18) results that a rigid body translation ${}_cU_i$ , implies local displacements ${}_iu_j = {}_iu_k = 0$ and the local forces ${}_nf_j$ and ${}_nf_k$ don't contribute to the total work of the system. But in the case of a deformable continuum there are modes of total deformation providing non-zero displacements of the points ${}_nr$ collinear with ${}_nf_j$ and ${}_nf_k$ and having the same direction with them. For this kind of motion, consisting of a mode of contra-formal deformation (Fig. 3.3) the corresponding expressions are:

$$ {}^c\zeta_{\bar{\imath}j} = {}^c\xi_{i\bar{j}} = \left(1-\delta_{ij}\right) r_i \hat{r}_j, \quad \left\{ {}^{ci}\zeta_{\bar{\imath}i} = {}^{ci}\xi_{i\bar{\imath}} = 0,\ {}^{ci}\zeta_{\bar{\imath}j} = {}^{ci}\xi_{i\bar{j}} = \left(r_i\hat{r}_j\right)^{\bar{k}} for\ i \neq j \right\} $$

(3.19a)

$$ {}^cF_i = \sum_j \sum_{n=1}^{8} {}^{ci}_{n}\xi_{i\bar{j}}\, {}_nf_j = \sum_j \sum_{n=1}^{8} \left( \left(1-\delta_{ij}\right)\left({}_nr_i\, {}_n\hat{r}_j\right)^{\bar{k}}\, {}_nf_j \right)_i = $$

$$ = \sum_{n=1}^{8} \left( {}_nr_i\left( {}_n\hat{r}_j\, {}_nf_j + {}_n\hat{r}_k\, {}_nf_k \right)^1 \right)_i = \left( r_i\left(\hat{r}_j f_j + \hat{r}_k f_k\right)^1 \right)_i \qquad (3.19b) $$

$$ u_i = {}^cU_i\ {}^c\zeta_{\bar{\imath}i} = 0, \qquad u_j = {}^cU_i\, {}^{ci}\zeta_{\bar{\imath}j} = \left(\left({}^cU_i r_i\right)^{\bar{1}} \hat{r}_j\right)_j, \{j \neq i\} $$

The motive ${}^{c}U_i$ represents the translation of the center of gravity of the initial rectangle after its deformation resulting to a trapezoidal shape, shown with dashed lines in (Fig. 3.3b)

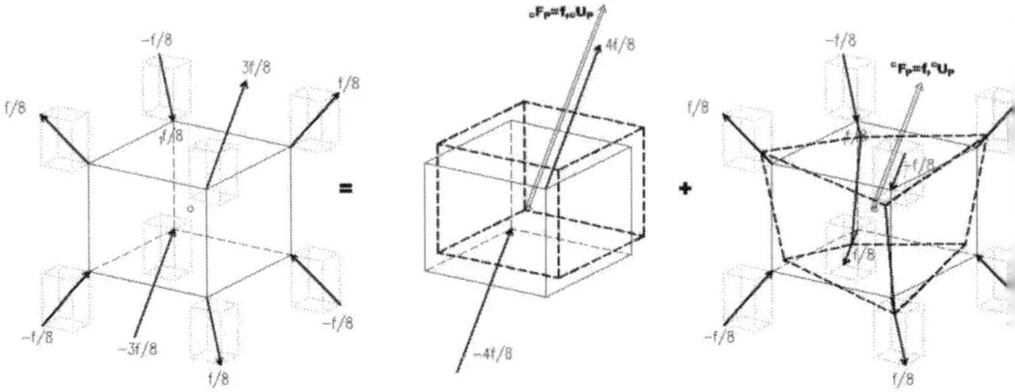

*Figure 3.4 - Conformal and Contra-formal Total Polar Vector Composants ${}_{c}F_p$ and ${}^{c}F_p$*

In the above relations, the composant ${}^{c}F_i$ and the motive ${}^{c}U_i$, both polar vectors defined on the $e_i$-axis, appear to correspond to local forces $f_j, f_k$ acting and displacements $u_j, u_k$ developed in the directions $e_j$ and $e_k$ perpendicular to the $e_i$-axis. A simple analogue of this state results after identifying the system of forces and displacements occurred, for example, on the plane $e^z$, with the system of the two cylinders in contact of a printer subjected opposite moments $\pm F^z/2 = \pm 2_n x\ {}_n f_y/4$, rotating thus counterwise and pushing a sheet of paper in their common tangential direction $x$ with a force ${}^{c}F_x$ at a translation ${}^{c}U_x$ (Fig. 3.4)

The motion ${}^{c}U_P$ and the total resultant ${}^{c}F_P$ for this translational deformation are represented by the polar vectors

$$ {}^{c}U_P = \sum_i {}^{c}U_i \qquad (3.19e) $$

$$ {}^{c}F_P = \sum_i {}^{c}F_i = \sum_i \left( (r_i \hat{r}_j) f_j + (r_i \hat{r}_k) f_k \right) \qquad (3.19g) $$

with

$$\xi_{P\bar{p}} = \zeta_{\bar{P}p} = \sum_i (r_i \hat{r}_j + r_i \hat{r}_k) = r \times \hat{r} = 0 \qquad (3.19h)$$

Which satisfies the invariance condition for ${}^cF_P$ and ${}^cU_P$.

According to the Eq. (3.18) the conformity condition (3.16c) takes the form

$$^{ci}u_{j,\bar{\imath}} = \left( {}^cU_i(r_i\hat{r}_j)\right)_{j,\bar{\imath}} = {}^cU_i\hat{r}_j \neq {}^{ci}u_{i,\bar{\jmath}} = 0 \qquad (3.20a)$$

$$^{Di}u_{j,\bar{k}} = {}^{Di}U_{k,\bar{\jmath}} = 0 \qquad (3.20b)$$

From these relations results that the motive ${}^cU_i$ is contra-formal in the direction of the $e_i$-axis and conformal on the plane $e^i$.

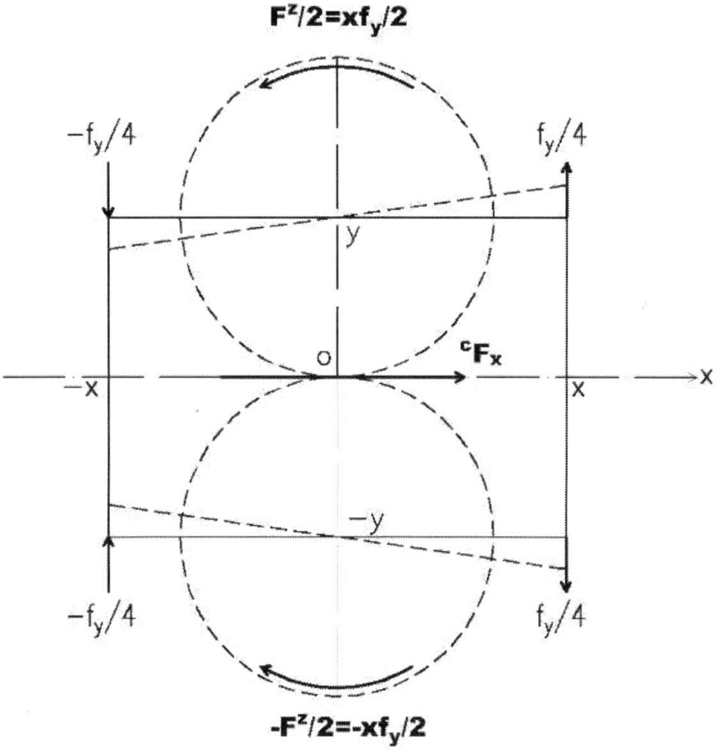


*Figure 3.5 - Contra-formal Polar Vectors in the Plane*

### 3.7 Axial Vector Composants and Motives

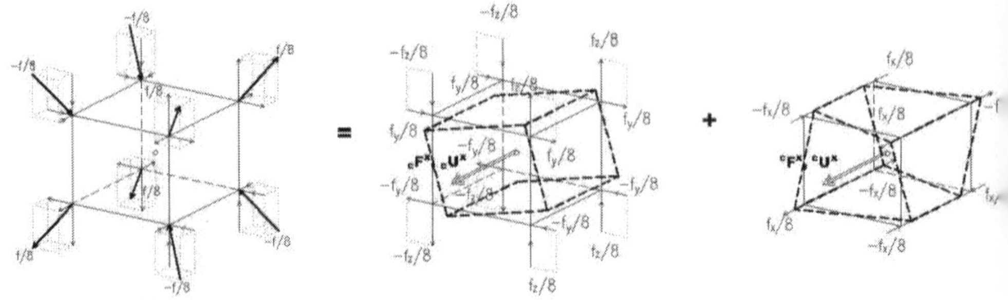

(e)

*Figure 3.6 - Conformal and Contra-formal Axial Vector Composants $_cF^x$ and $^cF^x$*

The system of forces shown in (Fig. 3.2e) is equivalent to a single moment composant with respect to the $e_x$ axis, which is represented by the relation

$$_cF^x = \sum_{n=1}^{8} \left( \left( {_nz} \ {_nf_y} \right)^{\bar{x}} + \left( {_ny} \ \ {_nf_z} \right)^{x} \right),$$

$${_n\xi^x}_{\bar{x}} = 0, \quad {_n\xi^x}_{\bar{y}} = {_nz}, \quad {_n\xi^x}_{\bar{z}} = {_ny},$$

For a rigid body and hence conformal rotation around the same axis, defined by the motive $_cU^x$, the local displacements of the points $_nr$ are

$${_nu_x} = 0, \quad {_nu_y} = \left( {_cU^x} \ {_nz} \right)_{\bar{y}}, \ {_nu_z} = \left( {_cU^x} \ {_ny} \right)_{z},$$

$${_n\zeta^{\bar{x}}}_x = 0, \quad {_n\zeta^{\bar{x}}}_y = {_nz}, \quad {_n\zeta^{\bar{x}}}_z = {_ny}$$

Similar to the above relations result for the system of forces shown in (Figs. 3.2f and g), which, for rigid body rotations around the axes $e^y$ and $e^z$, are equivalent to composants and motives $_cF^y$, $_cU^y$ and $_cF^z$, $_cU^z$ respectively, expressed by the following general relations concerning :
the fundamental shape and lever-arm functions

$${_c^i\zeta^{\bar{i}}}_j = {_c^i\xi^i}_j = (1 - \delta_{ij}) r_k,$$

or

$${}^{i}_{c}\zeta^{\bar{\imath}}{}_{i} = {}^{i}_{c}\xi^{i}{}_{i} = 0, {}^{i}_{c}\zeta^{\bar{\imath}}{}_{i} = {}^{i}_{c}\xi^{i}{}_{\bar{\jmath}} = r_k \text{ for } i \neq j \tag{3.21a}$$

the composants

$${}_{c}F^{i} = \sum_{j}\sum_{n=1}^{8} {}_{c}\xi^{i}{}_{\bar{\jmath}} f_j = \sum_{n=1}^{8}\left(\left({}_{n}r_j\, {}_{n}f_k\right)^{i} + \left({}_{n}r_k\, {}_{n}f_j\right)^{\bar{\imath}}\right) = r_j f_k + r_k f_j \tag{3.21b}$$

and the displacements of any point $r'$ for the motives ${}_{c}U^{i}$

$${}^{i}_{c}u_i = \sum_{j} {}_{c}U^{j}\zeta^{\bar{\jmath}}{}_{i} = \left({}_{c}U^{j}r'_{k}\right)_{i} + \left({}_{c}U^{k}r_{,j}\right)_{\bar{\imath}} \tag{3.21c}$$

The total resultant moment of the system is

$${}_{c}F^{A} = \sum_{i} {}_{c}F^{i} = \sum_{i}\sum_{n=1}^{8}\left({}_{n}r_j\ {}_{n}f_k + {}_{n}r_k\, {}_{n}f_j\right)^{i} = \sum_{n=1}^{8}\left({}_{n}r \times {}_{n}f\right)^{A} =$$

$$= \sum_{n=1}^{8}\left({}_{n}r_{n}f_t\right)^{A}$$

or

$${}_{c}F^{A} = r \times f = r f_t \tag{3.21d}$$

where $f_t$ is the transversal component of $f$ which is collinear with the characteristic vector $t$ in the polar system of the pair of vectors $r$ and $f$ according to the Eq. (1.2.54) in the paragraph 1.2.9 (Fig 1.2.4).

From the above relations results that a rigid body rotation around the $e_i$-axis implies non zero translations in the directions of the axes $e_j$ and $e_k$ and zero translation in the $e_i$direction. Moreover to the conjugate composant moment ${}_{c}F^{i}$ contributes only the components ${}_{n}f_j$ and ${}_{n}f_k$ of the forces ${}_{n}f$ without any contribution of the collinear components ${}_{n}f_i$. To the total resultant moment ${}_{c}F^{A}$ contributes only the components ${}_{n}f_t$ of the forces ${}_{n}f$ transversal to the direction of the position vector ${}_{n}r$, with zero contribution of the components ${}_{n}f_r$ collinear with ${}_{n}r$ .

Nevertheless in the case of a deformable body there are modes of deformation equivalent to rotation about an $e^i$-axis, represented by the motive ${}^{c}U^{i}$, implying additional translations ${}^{ci}u_i$ in the same direction. Moreover to the composant moment ${}^{c}F^{i}$, conjugate to this motive, contributes and the components ${}_{n}f_i$ of the forces ${}_{n}f$ collinear with the translations ${}^{a}u_i$ . The total resultant

$$ {}^{c}F^{A} = \sum_{i} {}^{c}F^{i} $$

will represent the effect of the component $f_r$ of $f$, which is neglected for the conformal motion. The corresponding relations result as follows.

The component $f_r$ of $f$ and the radius t from the pole ${}_{z}P$ corresponding to the position r are

$$ f_r = (r \cdot f)_1 \hat{r}, \qquad t = \sqrt{2}(re^s)_{\bar{t}} \tag{3.22} $$

where $e^s = \frac{r \times f}{|r \times f|}$

The total resultant moment ${}^{c}F^{A}$ of $f_r$ with respect to the pole of the polar system (Fig. 1.1.5), is

$$ {}^{c}F^{A} = \sum_{i} {}^{c}F^{i} = (t \times f) = tf_r = \sqrt{2}(re^s)_{\bar{t}}\hat{r}(r \cdot f)_1 = \sqrt{2}((re^s)_{\bar{t}} f_r)^s \tag{3.23a} $$

with composants

$$ {}^{c}F^{i} = \sqrt{2}e^{i}(\bar{e^{i}} \cdot ((re^s)_{\bar{t}} f_r)^s)_1 \tag{3.23b} $$

which are invariant under orthogonal transformations.

In Fig 3.6 is repeated the Fig 3.2.e1 corresponding to the resultant moment $F^A$ now referred to the polar system of the pair of vectors $r$ and $f = f_r + f_t$ instead of the global reference system *Oxyz* of the particle, which is considered as superposition of the system of ${}_{n}f_{t}$ corresponding to ${}_{c}F^{A}$ (Fig 3.6a) and of the system of ${}_{n}f_{r}$ corresponding to ${}^{c}F^{A}$ (Fig 3.6b). From Fig 3.7b becomes apparent that the moment ${}^{c}F^{A}$ equals actually the sum of two equal moments ${}^{c}F^{1/2}$ one of the forces ${}_{n}f_{t}$ for *n*=1 to 4 applied at the nodes of the upper face of the cubic mesh with respect to their common pole ${}_{z}P$, which coincides with the pole of the pair *r* and *f*, and a second one of the forces ${}_{n}f_{t}$ for *n*=5 to 8 applied at the nodes on the face lower of the cubic mesh with respect to their common pole ${}_{z}P$, which is the mirror point of ${}_{z}P$ with respect to the plane $e^{z}$ and constitutes a dipole moment.

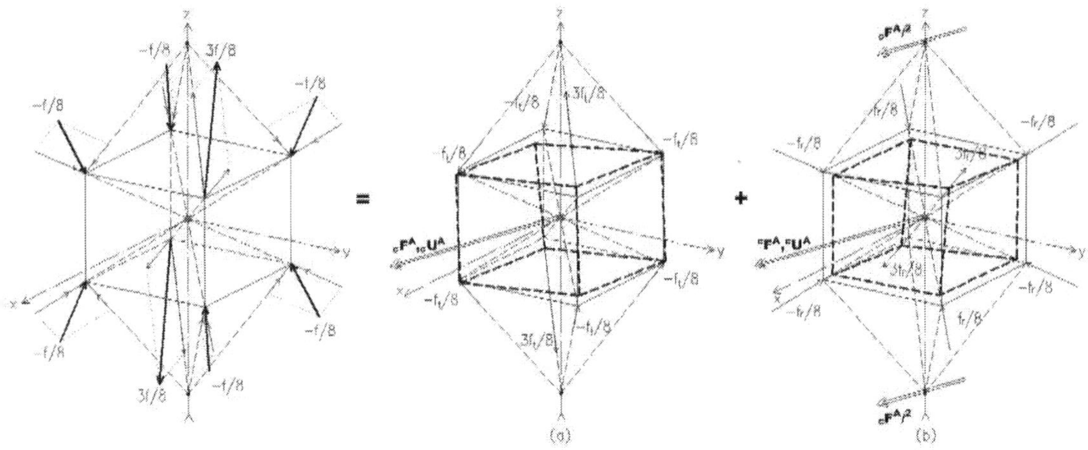


*Figure 3.7 - Total Axial Vector Composants and Motives in the Polar System*

The displacements of any point $r'$ if the particle for the motive ${}^{c}U^{A}$ are

$${}_{c}A_{U} = \sqrt{2}\ {}^{c}\quad{}^{A}\,(r'e^{s'}) \tag{3.24a}$$

and

$${}^{cA}u_{i} = \sqrt{2}\left(\left({}^{c}U^{A}(r'e^{s'})\right)\cdot \bar{e}_{i}\right)e_{i} \tag{3.24b}$$

where $e^{s'} = \frac{r\prime \times f}{|r\prime \times f|}$. From the obvious relation

$${}^{cA}u_{i,\bar{j}} \neq {}^{cA}u_{j,\bar{i}} \tag{3.25}$$

results that the motive is ${}^{c}U^{A}$ contra-formal

From Eq. (3.25) results that the motive ${}^{c}U^{A}$ is contra-formal.

### 3.8 Induction Moments

In algebra of octonions the effect on the origin of the system of polar vectors $f_p$ applied on resultant axial vectors, is not restricted exclussively to their cross products $(r \times f_p)^q$ and $(t \times f_p)^q$. According to Eqs. (1.2.7a), it contains the additional contribution of the product $(a_e f_p)^p$ of $f_p$ by an imaginary scalar $a_e$ , which results to an axial vector having the direction of $f_p$. Entities of this kind are not contained in the subject of continuum mechanics in its present state. Nevertheless such an effect appears in electrodynamics under the form of the electromagnetic induction. An electric current along a straight conductor implies the development of a magnetic field with cyclic trajectories around and normal to the conductor, represented by a magnetic moment. Inversely, a magnetic flux variable with time between two poles implies the development of electric currents along cyclic conductors normal to the direction of the magnetic flux. This effect is not related with systems of material points and could be the subject of an independent investigation.

### 3.9 Real Scalar Composants and Motives

The system of forces shown in (Fig. 3.2a) results to zero polar and axial vector resultant and the total work performed by them for motions, represented by polar or axial vectors, translations or rotations, equals likewise zero. But in the case of a deformable continuum there are modes of deformation which imply local displacements having the direction of the forces and thus a non-zero work of the forces is performed for these displacements. In the followings is shown that this work due to deformation equals the work of a resultant represented by a real scalar entity for a motion defining the mode of deformation, which is represented likewise by a real scalar entity.

The real scalar composants and motives should have a unique value in all reference systems. As appropriate reference systems are received the polar systems of the pairs of vectors ${}_nr$ and ${}_nf$ (Fig 3.7), where the forces ${}_nf$ are analyzed to their components ${}_nf_r$ and ${}_nf_t$ instead of the components ${}_nf_x$, ${}_nf_y$ and ${}_nf_r$ (Fig 3.8a) and one consisting of the forces $f_t$ (Fig 3.8b).

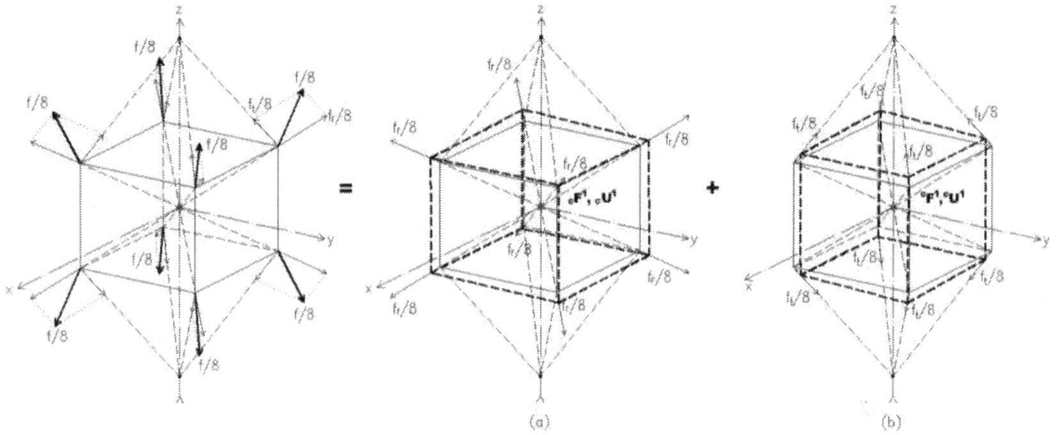

*Figure 3.8 - Real Scalar Composants and Motives in the Polar System*

The real scalar composant corresponding to the system of forces ${}_nf_n$ is defined by the expression

$$ {}_cF_1 = -\sum_{n=1}^{8}\left({}_nr\;{}_nf_r\right)_{\bar{1}} = -\sum_{n=1}^{8}\left({}_nr\cdot{}_nf\right)_{\bar{1}} = -\left(r\cdot f\right)_{\bar{1}} = -\sum_{i}\left(r_i f_i\right)_{\bar{1}} \qquad (3.26a)$$

which is invariant under orthogonal transformations, with fundamental lever-arm and shape functions

$$ {}_c\xi_{1_{\bar{i}}} = {}_c\zeta_{\bar{1}_i} = -r_i, \qquad \xi_{\bar{1}_{\bar{r}}} = \zeta_{\bar{1}_r} = -r \qquad (3.26b)$$

The displacements of any point $r'$ for the corresponding motive ${}_cU^1$ are

$$ {}_{c1}u_i = {}_cU_1\zeta_{\bar{1}_i} = {}_cU_1 r'_i, \qquad \sum_i {}_cu_i = {}_cU_1\sum_i r'_i = {}_cU_1 r' = {}_{c1}u_{r'} \qquad (3.26c)$$

In the polar system is $|r_i| = |r|/\sqrt{3}$ for any $i$ and Eq. (3.25c) represent a uniform dilatation of the particle in all directions. Furthermore it is

$$ {}_c u_{i,\bar{j}} = {}_c u_{j,\bar{i}} = 0 \tag{3.26d} $$

and the motion defined by the motive ${}_c U^1$ is conformal.

The real scalar composant corresponding to the system of forces ${}_n f_t$ is defined by the expression

$$ {}^c F_1 = \sum_{n=1}^{8} \left( {}_n t \; {}_n f_t \right)_{\bar{1}} = \sum_{n=1}^{8} \left( {}_n t \cdot {}_n f \right)_{\bar{1}} = (t \cdot f)_{\bar{1}} = \left( t t_f \right)_{\bar{1}} \tag{3.27a} $$

For $t = \sqrt{2}(re^s)_{\bar{t}}$ Eq. (3.27a) becomes

$$ {}^c F_1 = \sqrt{2}(re^s)_{\bar{t}} \cdot f = \sqrt{2}(re^s)_{\bar{t}} f_t \tag{3.27b} $$

which is invariant under orthogonal transformations, with fundamental lever-arm and shape functions

$$ {}^c \xi_{1_{\bar{t}}} = {}^c \zeta_{\bar{1}_t} = \sqrt{2}(re^s)_{\bar{t}} \tag{3.27c} $$

The displacements of any point $r'$ for the corresponding motive ${}^c U^1$ are

$$ {}^c_1 u = {}^c U_1 \, {}^c \zeta_{\bar{1}_t} = \sqrt{2} \, {}^c U_1 (r' e^{s'}) \tag{3.27d} $$

where $e^{s'} = \dfrac{r' \times f_p}{|r'||f_p|}$

From Eq. (3.27d) results that the displacements of the nodes are developed in the direction of the vectors ${}_n t$ and represent a dilatation on the planes parallel to the plane $e^s$ of the polar system. Furthermore it is

$$ {}^c_1 u_i = \sqrt{2} \, {}^c U_1 (\hat{r}'_i \cdot (r' e^{s'}) r'_i) \tag{3.27e} $$

$$ {}^c u_{i,\bar{j}} \neq {}^c u_{j,\bar{i}} \tag{3.27f} $$

and the motion defined by the motive ${}^c U_1$ is contra-formal.

## 3.10 Imaginary Scalar Composants and Motives

From the eight systems of forces shown in (Fig. 3.2), derived from the octal decomposition of the single force $f_p$ applied at the point ${}_1r$, the last remaining to be represented by a composant is that shown in (Fig. 3.2h). It is evident that this system results to zero polar and axial vector resultants as well as to zero real scalar composant. The remaining possibility is this system to be represented by imaginary scalar entities. The possibility the forces of the system to perform a work is restricted to the execution of displacements at their directions belonging to a mode of deformation. According to the procedure applied in the previous section, in the cases of displacements due to deformation of the continuum, will be used the representation of the system on the polar system of the pair of vectors ${}_1r$ and $f$, which is shown in (Fig. 3.8), where each one force, ${}_nf$, for $n = 1$ to 8, is analysed to the two components ${}_nf_r$ and ${}_nf_t$.

The system of the forces ${}_nf_r$ (Fig. 3.8a) is represented by the imaginary scalar composant

$$ {}_cF^e = \sum_{n=1}^{8} \left( {}_nr^a \, {}_nf_r \right)_e = (r^a f_r)^e = r^a (f \cdot r)_{\bar{1}} \hat{r} \qquad (3.28a) $$

where $r^a$ is the area vector of the position vector $r$.
After replacing $r^a$ with its expression (1.2.45a), Eq. (3.28a) takes the form

$$ {}_cF^e = \left( -\frac{|r|}{\sqrt{3}} e (r \cdot f)_{\bar{1}} \right)^e = -\frac{|r|}{\sqrt{3}} e \sum_i (r_i f_i)_{\bar{1}} = \left( -\frac{|r|}{\sqrt{3}} e (r f_r)_{\bar{1}} \right)^e \qquad (3.28b) $$

which is invariant under orthogonal transformations.

The fundamental lever-arm and shape functions are

$$ {}_c\xi^e{}_{\bar{\imath}} = {}_c\zeta^{\bar{e}}{}_i = -\frac{|r|}{\sqrt{3}} (e r_i)^{\bar{\imath}}, \qquad {}_c\xi^e{}_{\bar{r}} = {}_c\zeta^{\bar{e}}{}_r = -\frac{|r|}{\sqrt{3}} (e r)^{\bar{r}} \qquad (3.28c) $$

and the displacements for the corresponding motive ${}_cU^e$ are

$$ {}_{c}^{e}u_{i} = {}_{c}U^{e}\ {}_{c}\zeta^{\bar{e}}{}_{i} = \left(-\tfrac{|r|}{\sqrt{3}}\left({}_{c}U^{e}e\right)_{\bar{1}}r_{i}\right)_{i} \tag{3.28d} $$

$$ {}_{c}^{e}u_{r} = {}_{c}U^{e}\ {}_{c}\zeta^{\bar{e}}{}_{r} = \left(-\tfrac{|r|}{\sqrt{3}}\left({}_{c}U^{e}e\right)_{\bar{1}}r\right)_{r} \tag{3.28e} $$

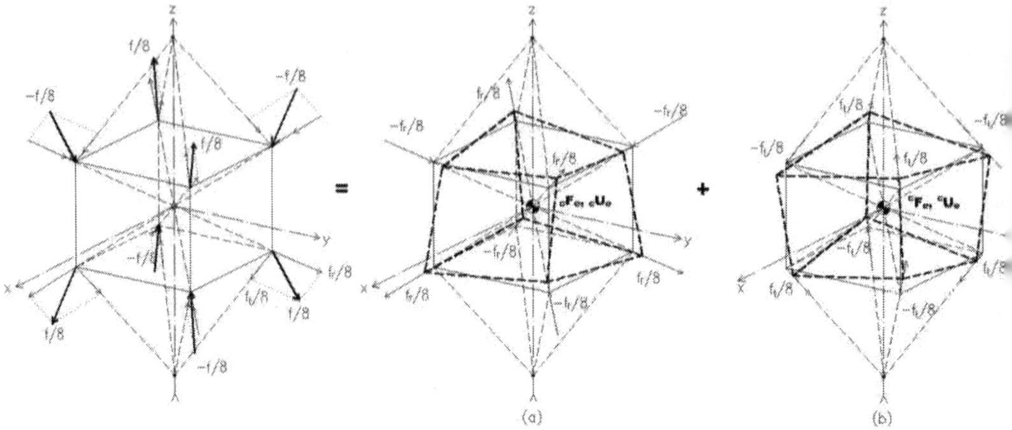

*Figure 3.9 - Imaginary Scalar Composants and Motives in the Polar System*

From the relation

$$ {}_{c}^{e}u_{i,\bar{j}} = \frac{r_{j}r_{i}}{\sqrt{3}|r|}\left({}_{c}U^{e}e\right)_{\bar{1}} = -\frac{r_{i}r_{j}}{\sqrt{3}|r|}\left({}_{c}U^{e}e\right)_{\bar{1}} = -{}_{c}^{e}u_{j,\bar{\imath}} \tag{3.28f} $$

results that the motive ${}_{c}U_{e}$ is conformal.

The system of the forces ${}_{n}f_{t}$ (Fig. 3.7b) is represented by the imaginary scalar composant

$$ {}^{c}F^{e} = -\sum_{n=1}^{8}\left({}_{n}t^{a}\ {}_{n}f_{t}\right)^{e} = -(t^{a}f_{t})^{e} = -t^{a}(f\cdot t)\hat{t} \tag{3.29a} $$

where $t^{a} = r^{a} + {}_{0}t^{a}$ is an area vector collinear with $t$ and ${}_{0}t^{a}$ the area vector of $t$. After replacing the vectors $t$ and $t^{a}$ with the expressions (1.2.44), (1.2.45b) the Eq. (3.30a) takes the form

$$ {}^{c}F^{e} =- t^{a}\hat{t}(t\cdot f)_{\bar{1}} = -\frac{|r|}{\sqrt{3}}(et)\hat{t}(t\cdot f) = \sqrt{\frac{2}{3}}|r|e((re^{s})_{\hat{t}}\cdot f)_{1} $$

$$ = \sqrt{\frac{2}{3}}|r|((e_{s}r)^{t}f_{t})^{e} \qquad (3.29b) $$

which is invariant under orthogonal transformations.

The fundamental lever-arm and shape functions are

$$ {}^{c}\xi^{e}{}_{\bar{t}} = {}^{c}\zeta^{\bar{e}}{}_{t} = \sqrt{\frac{2}{3}}|r|(e_{s}r)^{t} \qquad (3.29c) $$

and the displacements of any point $r'$ of the particle for the corresponding motive ${}^{c}U^{e}$ are

$$ {}^{ce}u_{t} = {}^{c}U^{e}\,{}^{c}\zeta^{e}{}_{t} = \sqrt{\frac{2}{3}}\,{}^{c}U^{e}|r'|e_{s\prime}r' \qquad (3.30a) $$

$$ {}^{ce}u_{i} = ({}^{ce}u_{t}\cdot\widehat{r'}_{i})r'_{i} = \sqrt{\frac{2}{3}}\,{}^{c}U^{e}|r'|\left((e_{s\prime}r')\cdot\widehat{r'}_{i}\right)r'_{i} \qquad (3.30b) $$

where $e_{s\prime} = e\left(\frac{r\prime\times f}{|r\prime\times f|}\right)$
From the obvious relation

$$ {}^{ce}u_{i,\bar{j}} \neq {}^{ce}u_{j,\bar{\iota}} \qquad (3.30c) $$

results that the motive ${}^{c}U^{e}$ is contra-formal.

In (Figs. 3.9a, b) the deformed shapes of the cubic mesh of the points ${}_{n}r$ for $n$=1 to 8 for the motives ${}_{c}U^{e}$ and ${}^{c}U^{e}$ are shown with dashed lines. They consist of displacements in the radial directions for ${}_{c}U^{e}$ and in directions transversal to $r$ and converging to the two poles of the polar systems of the pairs of vectors ${}_{n}r$ and ${}_{n}f$ for ${}^{c}U^{e}$. These modes of deformation corresponding to ${}_{c}U^{e}$ and ${}^{c}U^{e}$ can be referred as *central* and *dipole warping* respectively.

### 3.11 Spin and Bi-moment

In Quantum Mechanics the spin is referred as an intrinsing angular momentum $s^i$ which constitutes a specific characteristic of the elementary particles. The intrinsing character is attributed to the zero size of the particles although for each one of them a radius is contained to the list of their constitutive properties. But this radius is not a material dimension and represents the distance at which scattering between colliding similar particles occurs. The spin is experimentally detected in particles lying in magnetic fields, where their axis of rotation is aligned with the direction of the magnetic induction $B_i$. Outside the magnetic field, the axis of rotation is assumed to be oriented in various but distinct directions, which are defined by the spin quantum numbers.

Although the nature of the spin as angular momentum is manifested for particles affected by magnetic fields, a different aspect for their nature outside fields is derived on the basis of the following remarks· the magnetic induction $B_i$ is a vector which represents a geometric quantity, a length variation with time, similar to the intensity of the gravitational field, which corresponds to acceleration. The effect of the interaction between a line and a force vector is represented by their product and results to a new force referred t another octonion component, according to the multiplication table for octonions. Therefore the spin angular momentum $s^i$, which is developed in particles affected by magnetic fields, has to satisfy the relation

$$s^i = B_i s^e$$

where $s^e$ represents a bimoment or an angular bimomenture which, which, under this aspect, constitutes the actual intrinsic nature of the spin under free conditions, uninfluenced of magnetic fields. For the free state, the term spin, hinting a rotation around a single axis, should be replaced by the term bi-spin, or better, tri-spin.

### 3.12 The Complete System of Composants and Motives

From the previous discussion resulted that the complete description of the state of a particle in terms of overall state parameters requires two sets of conjugate pairs of composants and motives, one consisting of conformal and the other of contra-formal entities. The conformal entities are characterized as central or monopole and the contra-formal as dipole. Each one set contains three polar vectors, three axial vectors, both in the directions of a global orthogonal reference system, which constitute the components of the total polar and axial vector composants respectively and moreover two scalar entities, the one real and the other imaginary. The total composants and motives $F$ and $U$ representing the set of forces applied on and displacements developed by the material points of the particle, are shown in Table 3.1. The same functions correspond to the similar composants and motives $P$ and $V$ representing momentums and velocities of the material of the material particles.

In Table 3.1 the composants and the displacements are represented by the fundamental forms of the shape and lever-arm functions. For any specific case of particle and set of forces applied on its material points they have to be multiplied by appropriate even functions of $r^2$ which don't alter their scalar or vector character. These functions, as well as the corresponding motive $U$ depend on the shape, the size, the material properties and the set of the forces and are estimated applying the methods of the physics of deformable bodies.

By the fundamental shape functions of the six conformal motives $_cU_i$ and $_cU^i(i = x, y, z)$ are described rigid body motions of the particle, translations and rotations respectively, without the contribution of its deformation. By the rest two conformal motives $_cU_1$ and $_cU^e$ and the totality of the eight contraformal motives $^cU_I$ are described motions corresponding to ten different modes of deformation. From the six rigid body motions result the changes in the position and the orientation of the particle in the space without any effect on its internal structure and for that reason they are considered as external dimensions of the phase and the configuration space of the particle. At the other side the ten modes of deformation imply exclusively changes of the shape and the size of the particle which are strongly depended on the deformability properties of its material and constitute the dominant characteristics of its internal structure. For that reason they are considered as *internal* dimensions of the phase and of the internal configuration shape. The set of the ten internal dimensions is compatible with the ten dimensions provided from the Superstrings Theory for the elementary particles and

**Table 3.1**

| | Real Scalar | Polar Vector | Axial Vector | Imaginary Scalar |
|---|---|---|---|---|
| **Conformal or Monopole Mode** | ${}_cF_1 = -\sum (r \cdot f) = = -\sum r f_r$<br>${}_{c1}u = {}_cU_1 r' = = {}_{c1}u_{r'}$ | ${}_cF_P = \sum f$<br>${}_cF_i = ( {}_cF_P \cdot \bar{e}_i ) e_i$<br>${}_{cp}u = {}_cU_P$ | ${}_cF^A = \sum r \times f = = \sum r F_t$<br>${}_cF^i = ( {}_cF^A \cdot \bar{e}^i ) e_i$<br>${}^A_c u = {}_cU^A r' = {}^A_c u_t$ | ${}_cF^e = \sum r^a f_r = = -\sum \frac{\lvert r \rvert}{\sqrt{3}} (r f_r) e$<br>${}^e_c u = -\frac{\lvert r \rvert}{\sqrt{3}} ( {}_cU^e e ) r'$ |
| **Contra-formal or Dipole Mode** | ${}^cF_1 = \sqrt{2} \sum (r e^s)_{\bar{t}} \cdot f = \sqrt{2} \sum (r e^s)_{\bar{t}} \cdot f_t$<br>${}^c_1 u = \sqrt{2}\, {}^cU_1 (r' e^{s'})$ | ${}^cF_P = \frac{1}{3\sqrt{2}} \sum e^e f$<br>${}^cF_i = ( {}^cF_P \cdot \bar{e}_i ) e_i$<br>${}^c_P u = \frac{1}{3\sqrt{2}}\, {}^cU_P e^{s'}$ | ${}^cF^A = \sum t \times f = \sum t f_r = = \sqrt{2} \sum (r e^s)_{\bar{t}} f_r$<br>${}^cF^i = ( {}^cF^A \cdot \bar{e}^i ) e^i$<br>${}^{cA}u = \sqrt{2}\, {}^cU^A (r' e^{s'})$ | ${}^cF^e = -\sum t^a f_t = = \sqrt{\frac{2}{3}} \sum \lvert r \rvert (e_s r) f_t$<br>${}^{ce}u = = \sqrt{\frac{2}{3}}\, {}^cU^e \lvert r' \rvert e_{s'} r'$ |
| **Auxiliary Relations** | $e^s = \frac{r \times f}{\lvert r \times f \rvert}, \quad e_s = e e^s, \quad t = \sqrt{2} r e^s, \quad f_r = (f \cdot \hat{r}) r, \quad f_t = \left(f \cdot (\widehat{r e^s})\right) (r e^s)$ | | | |

possibly they offer an idea for the geometrical and mechanical meaning for these dimensions.

The six conformal composants ${}_cF_i$, ${}_cF^i$ or ${}_cP_i$, ${}_cP^i$ $(i = x, y, z)$corresponding to the external dimensions of the particle are identical to the non zero polar or axial vector sum of the forces or the momentums respectively acting on its material points and are identified with the traditional entity of the polar or axial vector resultant. For the rest ten composants consisting of the two conformal scalar ${}_cF_1$, ${}_cF^e$or ${}_cP_1$, ${}_cP^e$ and the eight contraformal scalar and vector ${}^cF_1$, ${}^cF_i$, ${}^cF^i$, ${}^cF^e$ or ${}^cP_1$, ${}^cP_i$, ${}^cP^i$, ${}^cP^e$, which correspond to the internal dimensions, the polar and the axial vector sums of the forces or the momentums acting on the material particles equal zero. The six motives ${}_cU_i$, ${}_cU^i$ or ${}_cV_i$, ${}_cV^i$ corresponding to the external composants represent rigid body displacements or motions of the particle, while the remaining ten, ${}_cU_1$ , ${}_cU^e$, ${}^cU_1$ , ${}^cU_i$ , ${}^cU^i$ , ${}_cU^e$ or ${}_cV_1$ , ${}_cV^e$, ${}^cV_1$ , ${}^cV_i$ , ${}^cV^i$, ${}^cV^e$ , corresponding to the internal composants represent displacements or motions due to the deformation. For both, the external and the internal pairs of conjugate entities $F_I, U_I$ or $P_I, V_I$, the work done or the kinetic energy aquired by the composants for their conjugate motive is equal to the sum of the work done or the kinetic energy acquired by the forces of the momentums acting on the material points of the particle for their displacements or velocities respectively. On the basis of this common characteristic property the composants and the motives constitute a generalization of the concept of the “resultant”, which is not restricted to the vector sum of the forces or momentums acting on, and of the displacements or velocities executed by the material points of the particle.

# 4. ONE-DIMENSIONAL AGGREGATES OF TWO-DIMENSIONAL PARTICLES

## 4.1. Beams and Strings

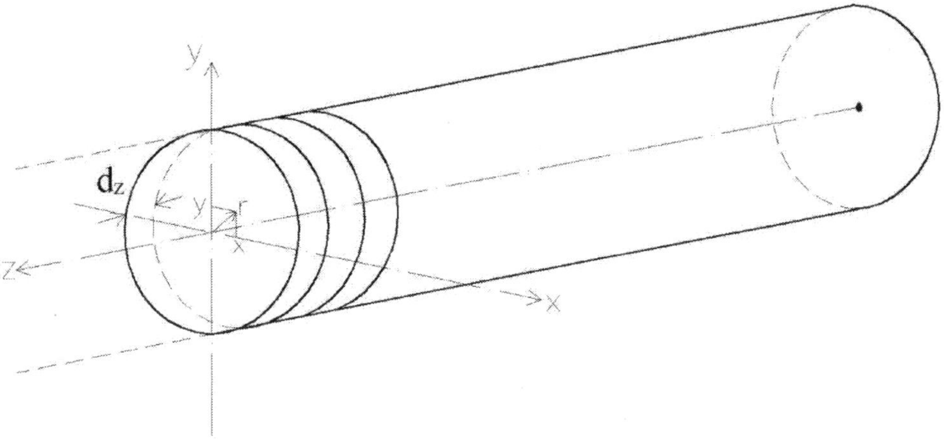


*Figure 4.1 - 2-D Particles of Beams*

As beams are considered continuum segments of a finite size having the dimension in one direction long enough compared to its dimensions in the directions transversal with respect to the longitudinal one. The line passing through the center of mass of the cross sections transversal to the longitudinal direction is the axis of the beams. Examples of beams are the linear solid members of structures, the rotating shafts of machines, the streams of fluids flowing in a narrow channel or in a tube etc.

The beam is considered as a one dimensional aggregate of particles consisting of slices of infinitesimal thickness between transversal sections (Fig. 4.1). The composants and motives of each one particle are the state parameters of the beam at the corresponding point of the axis. The continuity condition along the axis requires equal displacements, velocities, strains and strain rates to be developed at the two faces of the material points lying on the interfaces between adjacent particles. This condition implies that equal stresses and impulses are likewise developed at these faces, which constitute the interacting forces and momentums between particles.

As the target of the present treatise is restricted to the definition of the composants and motives, which constitute the state parameters for this specific type of particles, the investigation is simplified by considering the static case of solid deformable beams. For that case the laminar state of the continuum is considered where only polar vector forces are applied and polar vector displacements are developed at the material points without the effect of momentums and velocities. The definitions for the composants and motives concerning momentums and velocities respectively, as for example for the case of beams representing streams of fluids, are similar to those resulting for forces and displacements.

The solid deformable beams constitute a main subject of Structural Mechanics. In its present formulation the state of the particles is represented by six composants of the interacting stresses at the interfaces and the corresponding to them six motives. Three of them are polar vectors, one axial force in the direction of the axis of the beam and two shear forces in two perpendicular directions on the plane of the interface, with the corresponding motives to represent rigid body translations along the same directions. The rest three are axial vectors, one torsional moment around the axis of the beam and two bending moments around the two perpendicular axes on the plane of the interface, with the corresponding motives to represent rigid body rotations around the same axes, one torsional and two flexural respectively. For a specific type of beams, those with thin wall tubular cross sections, have been found that the effect of a seventh composant concerning the normal interacting stresses developed at the interfaces between adjacent particles and known as *bi-moment*, is decisive for their behavior, although this effect on beams with compact cross sections is negligible. From the following investigation results that the bi-moment is expressed by a triple polar vector product implying that in the octonion formulation it is represented by an imaginary scalar entity. Moreover from this investigation results that to the shear stresses applied at the interfaces corresponds an eighth composant represented by a real scalar entity for which the term *diverging shear* or shortly *di-shear* is proposed. The conjugate to the bi-moment and the di-shear represent modes of deformation and not rigid body motions. The eight pairs of composants and motives are elements of a pair of octonions and correspond to conformal motions.

The work done by the above composants for the corresponding motives do not exhaust the total work done by the forces applied on the material points of the particle for the corresponding local displacements. From the following investigation results that to the interacting stresses corresponds another set of eight pairs of contra-formal composants and motives which represent exclusively modes of deformation of the particle. Moreover, except the interacting stresses developed at the interfaces, there are stresses developed at the interfaces between material points contained in the interior of the particle between the two end faces. It is proved that they constitute two more sets each one consisting of eight composants and motives, one conformal and one contra-formal. They will be referred as *Hidden* composants and motives and they correspond exclusively to modes of deformation. For the solid beams consisting of a stiff material with low deformability the work done by the 26 pairs of composants and motives corresponding to modes of deformation is negligible compared with that done by the 6 pairs corresponding to rigid body motions. For that reason they are neglected in the running version of Structural Mechanics which is oriented to the analysis of structures contained in the everyday construction practice. But for beams consisting of a flexible material with high deformability, as for example the flowing streams of liquids and gases, where the kinetic energy of the material points is predominant, due to high values of momentums, velocities, impulses and strain rates, the effect of the state parameters which correspond to modes of deformation is significant and cannot be neglected.

From the 16 pairs of interacting state parameters the 6 correspond to rigid body motions and the 10 to modes of deformation. Furthermore from the total number of the 32 state parameters, interacting and hidden, the same set of 6 corresponds to rigid body motions, and the rest 26 to modes of deformation. The 6 rigid body motions define changes in the position and the orientation of the particle in the 3 dimensional space and they are referred to external dimensions of its configuration and phase spaces. At the other side the modes of deformation depend of properties characteristic of the internal structure of the particle and hence they are refereed to internal dimensions of the configuration and phase spaces. The 10 internal dimensions of the interacting state parameters and the 26 internal dimensions of the total set of state parameters, interacting and hidden, appear an important similarity with the 10 dimensions provided by the Theories of Super Strings and the 26 dimensions provided by the M-theory for the elementary particles.
Generally the axis of a beam is a continuous curve in the space with the shape and the size of the cross sections varying along it. The analytical treatment of the state of beams for this general geometric form is accessible

in cases where the curve of the axis and the variation of the geometrical properties of the cross sections are defined by normal analytical expressions. Nevertheless by using the recently developed Finite Element Method the treatment of the state of beams of any form becomes accessible. The axis is replaced by a polygonal line consisting of straight segment between the node points, which is inscribed in the curve. The geometrical properties of the cross sections along each one straight segment have constant values equal to the average values of these properties along the corresponding segment of the curve. The desired accuracy is obtained by choosing the appropriate density of the nodes on the curve. Under this view the further investigation can be limited to the simpler case of beams with straight axis and constant cross section along its length without loss of the main target which consists to the definition of the state parameters.

## 4.2. Reference Systems

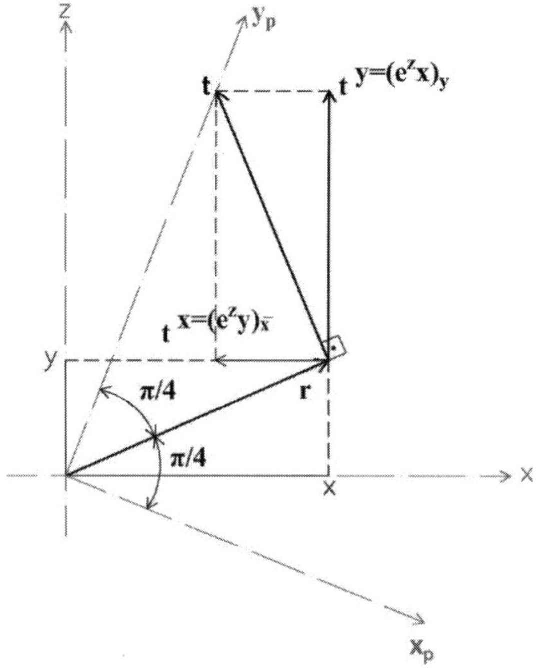


*Figure 4.2 - Polar Coordinates*

The orthogonal global reference system (Fig. 4.2) has the origin at the center of mass of the cross section. The axis $z$ coincides with the axis of the beam and the axes $x$ and $y$ lye on the plane of the cross section which is transversal to the $z$-axis. The coordinate $z$ defines the position of the particles along the axis of the beam. The position vector of a material points with elementary area on the cross section $dA^z = (dxdy)^z$ and elementary volume $dV_e = (dA^z dz)_e$ as well as its area vector in the reference system of each one particle, which is a transversal slice of infinitesimal thickness $dz$, are defined by the expressions

$$r = x + y, \qquad r^a = (xy)^z \tag{4.1}$$

The polar vector forces $f$ applied at the material points are considered consisting of two components, one lateral $f_l$ and one axial $f_z$

$$f = f_l + f_z, \qquad f_l = f_x + f_y \tag{4.2}$$

The polar systems are defined for the pair of vectors $r$ and $f_l$, both lying on the plane $oxy$ and hence it is $e^q = e^z$, which implies that the polar system corresponding to a position vector $r$ is unique for all pairs, $f_l$ . The pole lies on the $y$-axis and the coordinates of the characteristic vectors in the polar system are (Fig. 4.2)

$$r = \frac{|r|}{\sqrt{2}}(e_x + e_y), \qquad t = \frac{|r|}{\sqrt{2}}(-e_x + e_y), \qquad |t|^2 = |r|^2 \tag{4.3}$$

with area vectors

$$r^a = -t^a = \frac{|r|^2}{2} e^z \tag{4.4}$$

For the points on the perimeter $S$ of the cross section local orthogonal reference systems $nsz$ are defined (Fig. 4.2). The $s$-axis coincides with the tangent to the perimeter at the point considered in the counterclockwise direction, the $n$-axis is normal to both $s$ and $z$ axes and the perimetric surface element of the corresponding material point is

$$dA^n = (dsdz)^n \tag{4.5}$$

### 4.3. The Bi-quaternal Decomposition

The definition of the kinds of composants and motives for the 3-dimensional particles is obtained through the introduction of the octal decomposition procedure which is presented in the previous chapter 3. For the 2-dimensional particles constituting the beams the above target is obtained by the introduction of the following bi-quaternal decomposition procedure. For that purpose the effect of a single force $f = f_x + f_y + f_z$ on the composants is likewise considered which is applied on a single mass point ${}_1r = {}_1x + {}_1y$ of the particle. Two separate quaternal decomposition procedures are applied, the one concerning the lateral components $f_l = f_x + f_y$ of $f$ and the other concerning the axial component $f_z$ . (Figs. 4.5 and 4.6).

At these two procedures the forces $f_l$ and $f_z$ are replaced by the superposition of four systems each one consisting of four forces which are applied on the four mirror points ${}_nr = \pm\, {}_1x \pm\, {}_1y$ of ${}_1r(n = 1\ to\ 4))$ with respect to the coordinate axes $x, y$ resulting for all possible compinations of the signs of ${}_1x$ and ${}_1y$ . At the point ${}_1r$ the forces ${}_1f_l = f_l/4$ or ${}_1f_z = f_z/4$ while at the rest three points the forces ${}_nf_A = \pm\, {}_1f_A\ /4$ or ${}_nf_z = \pm\, {}_1f_z\ /4$ are applied. For ${}_nf_l$ the plus sign denotes the orientation similar to that of $f_l$ with respect to the origin of the reference system. In one of the four systems the four forces ${}_nf$ receive the plus sign. In the rest three systems the force ${}_1f$ and one of the forces for $n \neq 1$ receive the plus while the rest two forces receive the minus sign. From the superposition of the four systems of forces zero resultants at the points ${}_nr$ for $n \neq 1$ are obtained while at the point ${}_1r$ the resultant is $f_l$ or $f_z$ for the two separate quaternal decomposition procedures respectively.

The above quaternal decomposition procedures are unique. This uniqueness the necessity implies for the acceptance of real and imaginary scalar, beyond the directly recognized polar and axial vector composants and motives, as well as their conjugates corresponding to contra-formal motions.

## 4.4. Lateral Force Composants and Motives

### 4.4.1. Polar Vector Composants and Motives

The system of forces shown in (Fig. 4.3b) is furthermore decomposed in the superposition of the two systems shown in (Figs. 4.3b1 and 4.3b2). To the first of them corresponds the composant

$$ {}_{c}F_{x} = \sum_{n=1}^{4} {}_{n}f_{x} = f_{x} = {}_{c}\xi_{x\bar{x}} f_{x} \tag{4.6}$$

with the fundamental lever-arm and shape functions

$$ {}_{c}\xi_{x\bar{x}} = {}_{c}\zeta_{\bar{x}x} = 1, \qquad {}_{c}\xi_{x\bar{y}} = {}_{c}\zeta_{\bar{x}y} = 0 \tag{4.6a}$$

For the corresponding motive ${}_{c}U_{x}$ the displacements of any point $r = x + y$ are

$$ {}_{c}u_{x} = {}_{c}U_{x}\, \zeta_{\bar{x}x} = {}_{c}U_{x} = constant, \qquad {}_{c}u_{y} = {}_{c}U_{x}\, {}_{c}\zeta_{\bar{x}y} = 0 \tag{4.6b}$$

which defines a rigid body translation of the particle in the x-direction and constitutes a conformal motion.

The forces of the second system shown in (Fig. 4.3b2) execute zero work for the rigid body motion ${}_{c}U_{x}$ . But in the case of a deformable body there is a mode of deformation which implies displacements in the direction of the forces ${}_{n}f_{y}$ and the work done by them should be equal to the work of a composant ${}^{c}F_{x}$ for the corresponding to it motive ${}^{c}U_{x}$ . From the model introduced in 3.6 and illustrated in (Fig. 3.4) results

$$ {}^{c}F_{x} = \sum_{n=1}^{4} \left( {}_{n}x_{n}\, \hat{y} \right) {}_{n}f_{y} = \left( (x\hat{y})\, {}^{\bar{z}}f_{y} \right)_{x} \tag{4.7}$$

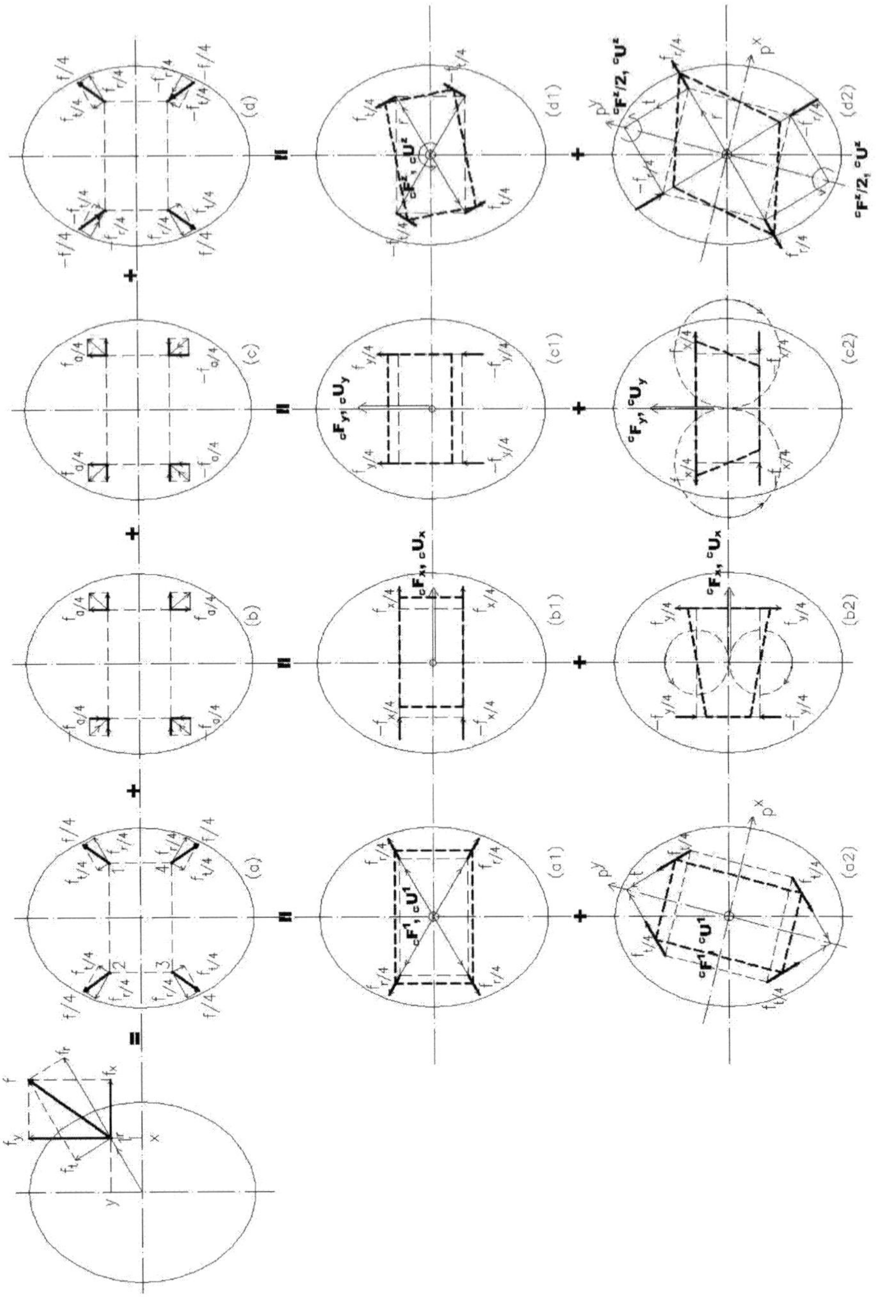
(a)
(b)
(c)
(d)
(a1)
(b1)
(c1)
(d1)
(a2)
(b2)
(c2)
(d2)

with the fundamental lever-arm and shape functions

$$ {}^{c}\xi_{x\bar{x}} = {}^{c}\zeta_{\bar{x}x} = 0, \qquad {}^{c}\xi_{x\bar{y}} = {}^{c}\zeta_{x\bar{y}} = (x\hat{y}) = \left(-\frac{xy}{|r|^2}\right)^{\bar{z}} \quad (4.7a) $$

For the motive ${}^{c}U_x$ the displacements of any point $r = x + y$ are

$$ {}^{c}u_x = {}^{c}U_x \; {}^{c}\zeta_{\bar{x}x} = 0, \qquad {}^{c}u_y = {}^{c}U_x \; {}^{c}\zeta_{\bar{x}y} = \left( {}^{c}U_x \, (x\hat{y})^{\bar{z}} \right)_y \quad (4.7b) $$

it is

$$ {}^{c}u_{x,\bar{y}} = 0, \qquad {}^{c}u_{y,\bar{x}} = -\,{}^{c}U_y \; \hat{y} \neq {}^{c}u_{z,\bar{y}} \quad (4.7c) $$

and the motion represented by the ${}^{c}U_x$ is contra-formal. For this mode of deformation the rectangle with summuts the points ${}_{n}r$ is transformed to a trapezoidal shape which is shown with dashed lines in (Fig. 4.3b1).

From similar to the above procedure result the following relations concerning polar vector composants and motives in the y-direction corresponding to the system of forces shown in (Figs. 4.3c, c1, c2), the conformal one

$$ {}_{c}F_y = {}_{c}\xi_{y\bar{y}} f_y = f_y, \quad (4.8) $$

With

$$ {}_{c}\xi_{y\bar{x}} = 0, \qquad {}_{c}\xi_{y\bar{y}} = \qquad 1 \quad (4.8a) $$

and the contra-formal

$$ {}^{c}F_y = {}^{c}\xi_{y\bar{x}} f_x = ((y\hat{x})^z f_x)_y \quad (4.9) $$

With

$$ {}^{c}\xi_{y\bar{x}} = y\hat{x} = -\frac{yx}{|r|}, \qquad {}^{c}\xi_{y\bar{y}} = 0 \quad (4.9a) $$

The total polar vector composants are

$$ {}_{c}F_P = {}_{c}F_x + {}_{c}F_y = f_x + f_y = f_l \quad (4.10) $$

$$^{C}F_{P} = {}^{C}F_{x} + {}^{C}F_{y} = -\frac{xy}{|r|^2}(f_x - f_y) \quad (4.11)$$

with

$$^{C}\xi_{p\bar{p}} = -\frac{xy}{|r|^2}(1 - 1) = 0 \quad (4.11a)$$

and the invariance condition under orthogonal transformations for the total composants ${}_{C}F_{P}$ and ${}^{C}F_{P}$ is satisfied. The composants ${}_{C}F_{x}$, ${}_{C}F_{y}$, ${}^{C}F_{x}$, ${}^{C}F_{y}$ are generally referred as *shear forces*.

### 4.4.2. Axial Vector Composants and Motives

The polar vector resultant for the system of forces shown in (Fig. 4.3d) is zero corresponds to it a non zero axial vector composant

$${}_{C}F^{Z} = \sum_{n=1}^{4}\left({}_{n}y_{n}f_{x} + {}_{n}x\,{}_{n}f_{y}\right)^{Z} = \left(yf_x + xf_y\right)^{Z} = r \times f = (rf_t)^{Z} \quad (4.12)$$

where $f_t$ is the component of $f$ transversal to the position vector $r$ .

with fundamental lever-arm and shape functions

$${}_{C}\xi^{Z}{}_{\bar{x}} = {}_{C}\zeta^{\bar{Z}}{}_{x} = y, \qquad {}_{C}\xi^{Z}{}_{\bar{y}} = \zeta^{\bar{Z}}{}_{y} = x \quad (4.12a)$$

$${}_{C}\xi^{Z}{}_{\bar{t}} = {}_{C}\zeta^{\bar{Z}}{}_{t} = r, \qquad {}_{C}\xi^{Z}{}_{r} = {}_{C}\zeta^{\bar{Z}}{}_{r} = 0 \quad (4.12b)$$

${}_{C}F^{Z}$ is invariant under orthogonal transformations of the system $oxy$ and is referred as *torsional moment.* For the corresponding motive ${}_{C}U^{Z}$ the displacements of any point $r = x + y$ are

$${}_{C}u_{x} = {}_{C}U^{Z}\zeta^{\bar{Z}}{}_{x} = {}_{C}U^{Z}y, \qquad {}_{C}u_{y} = {}_{C}U^{Z}\zeta^{\bar{Z}}{}_{y} = {}_{C}U^{Z}x \quad (4.12c)$$

$${}_{C}u_{r} = {}_{C}U^{Z}\zeta^{\bar{Z}}{}_{r} = 0, \qquad {}_{C}u_{t} = {}_{C}U^{Z}\zeta^{\bar{Z}}{}_{t} = \left({}_{C}U^{Z}r\right)_{t} \quad (4.12d)$$

it is

$$ {}_{c}u_{x,\bar{y}} = {}_{c}u_{y,\bar{x}} = {}_{c}U^{z} \tag{4.12e}$$

and implies that the motion represented by the motive ${}_{c}U^{z}$ is conformal consisting of a rigid body rotation around the z-axis.

For the rigid body rotation ${}_{c}U^{z}$ implying displacements in the direction $t$, which is transversal to the position vector $r$, the component $f_r$ of $f$ executes zero work. But in the case of a deformable body there is a mode of deformation, represented by an axial vector motive ${}^{c}U^{z}$ which provides displacements in the $r$-direction due to rotations around the $z$-axis at the two poles of the polar system of the position vector $r$. As ${}^{c}F^{z}$ have to be invariant under orthogonal transformations in Fig 4.3d2 the forces ${}_{n}f_{z}$ are referred to the polar system of $r$, which corresponds to the case the quaternal decomposition is initially applied to that system. The corresponding dipole composant is

$$ {}^{c}F^{z} = -\sum_{n=1}^{n} {}_{n}t\, {}_{n}f_{r} = -tf_{r} = -t \times f = -(e^{z}r)_{t} \times f = (e^{z}x)_{\bar{y}}f_{x} - (e^{z}y)_{x}f_{y} \tag{4.13}$$

with fundamental lever-arm and shape functions

$$ {}^{c}\xi^{z}{}_{\bar{x}} = {}^{c}\zeta^{\bar{z}}{}_{x} = -e^{z}x, \qquad {}^{c}\xi^{z}{}_{\bar{y}} = {}^{c}\zeta^{\bar{z}}{}_{y} = -e^{z}y \tag{4.13a}$$

$$ {}^{c}\xi^{z}{}_{\bar{r}} = {}^{c}\zeta^{\bar{z}}{}_{r} = -t = -e^{z}r, \qquad {}^{c}\zeta^{z}{}_{\bar{t}} = {}^{c}\zeta^{\bar{z}}{}_{t} = 0 \tag{4.13b}$$

which is likewise invariant under orthogonal transformations and which is referred as *distortional moment.*

For the corresponding motive ${}^{c}U^{z}$ the displacements of any point $r = x + y$ are

$$ {}^{c}u_{x} = {}^{c}U^{z}\, {}^{c}\zeta^{\bar{z}}{}_{x} = -\,{}^{c}U^{z}(e^{z}x)_{\bar{y}}, \qquad {}^{c}u_{y} = {}^{c}U^{z}\, {}^{c}\zeta^{\bar{z}}{}_{y} = -\,{}^{c}U^{z}(e^{z}y)_{x} \tag{4.13c}$$

$$ {}^{c}u_{r} = {}^{c}U^{z}\, {}^{c}\zeta^{\bar{z}}{}_{r} = -\,{}^{c}U^{z}(e^{z}r)_{t}, \qquad {}^{c}u_{t} = {}^{c}U^{z}\, {}^{c}\zeta^{\bar{z}}{}_{t} = 0 \tag{4.13d}$$

it is

$$ {}^{c}u_{x,\bar{y}} = - {}^{c}U^{z}e^{z} \neq - {}^{c}u_{y,x} = {}^{c}U^{z}e^{z} \tag{4.13e} $$

$$ {}^{c}u_{r,\bar{t}} = - {}^{c}U^{z} \neq - {}^{c}u_{t,\bar{r}} = 0 \tag{4.13f} $$

and the motion represented by the motive ${}^{c}U^{z}$ is contraformal, and consists of two unidirectional rotations around the two poles of the polar system of $r$. In (Fig. 4.3d2) with thick dashed lines is shown the rhomboidal shape of the deformed square which has summits the points ${}_{n}r$ in the polar system of $r$.

### 4.4.3. Real Scalar Composants and Motives

The system of forces shown in (Fig. 4.3a) corresponds to zero polar and axial vector composants and the work done by these forces for displacements due to a mode of deformation have to be represented by the work of a real scalar composant for a similar motive. The real scalar should have a constant value at all orthogonal reference systems including the polar systems of the position vectors ${}_{n}r$. At these last systems the following two real scalar expressions for the composants are offered.

The first expression is

$$ {}_{c}F^{1} = -\sum_{n=1}^{4} {}_{n}r \cdot {}_{n}f = -r \cdot f = -rf_{r} = -\left(xf_{x} + yf_{y}\right)^{\bar{1}} \tag{4.14} $$

with fundamental lever-arm and shape functions

$$ {}_{c}\xi^{1}{}_{\bar{x}} = {}_{c}\zeta^{\bar{1}}{}_{x} = -x, \qquad {}_{x}\xi^{1}{}_{\bar{y}} = {}_{c}\zeta^{\bar{1}}{}_{y} = -y \tag{4.14a} $$

$$ {}_{c}\xi^{1}{}_{\bar{r}} = {}_{c}\zeta^{\bar{1}}{}_{r} = -r, \qquad {}_{c}\xi^{1}{}_{\bar{t}} = {}_{c}\zeta^{\bar{1}}{}_{t} = 0 \tag{4.14b} $$

The displacements for the corresponding motive ${}_{c}U^{1}$ are

$$ {}_{c}u_{x} = {}_{c}U^{1}\,{}_{c}\zeta^{\bar{1}}{}_{x} = -\,{}_{c}U^{1}x, \qquad {}_{c}u_{y} = {}_{c}U^{1}\,{}_{c}\zeta^{\bar{1}}{}_{y} = -\,{}_{c}U^{1}y \tag{4.14c} $$

$$ {}_{c}u_{r} = {}_{c}U^{1}\,{}_{c}\zeta^{\bar{1}}{}_{r} = -\,{}_{c}U^{1}r, \qquad {}_{c}u_{t} = {}_{c}U^{1}\,{}_{c}\zeta^{\bar{1}}{}_{t} = 0 \tag{4.14d} $$

It is

$$ {}_c u_{x,\bar{y}} = -\ {}_c u_{y,\bar{x}} = 0 \tag{4.14e}$$

$$ {}_c u_{r,\bar{t}} = -\ {}_c u_{t,\bar{r}} = 0 \tag{4.14f}$$

From Eqs. (4.14e, f) results that the motion denoted by the motive ${}_cU^1$ is conformal and represents a mode of deformation consisting of a dilatation on the $oxy$ plane which is uniform along the $r$-directions (Fig. 4.3a1).

The second real scalar composant is

$$ {}^cF^1 = -\sum_{n-1}^{4} {}_nt \cdot {}_nf = -t \cdot f = -tf_t = -(e^z r)_t \cdot f = -\big((e^z y)_{\bar{x}} f_x + (e^z x)_y f_y\big) \tag{4.15}$$

where $t = e^z r = e^z(x+y) = (e^z y)_{\bar{x}} + (e^z x)_y$ with fundamental lever-arm and shape functions

$$ {}^c\xi^1{}_{\bar{x}} = {}^c\zeta^{\bar{1}}{}_x = -e^z y, \qquad {}^c\xi^1{}_{\bar{y}} = {}^c\zeta^{\bar{1}}{}_y = -e^z x \tag{4.15a}$$

$$ {}^c\xi^1{}_{\bar{r}} = {}^c\zeta^{\bar{1}}{}_r = 0, \qquad {}^c\xi^1{}_{\bar{t}} = {}^c\zeta^{\bar{1}}{}_{\bar{t}} = -t = -e^z r \tag{4.15b}$$

The displacements for the corresponding motive ${}^cU^1$ are

$$ {}^cu_x = {}^cU^1\, {}^c\zeta^{\bar{1}}{}_x = -\ {}^cU^1 (e^z y)_{\bar{x}}, \qquad {}^cu_y = {}^cU^1\, {}^c\zeta^{\bar{1}}{}_y = -\ {}^cU^1 (e^z x)_y \tag{4.15c}$$

$$ {}^cu_r = {}^cU^1\, {}^c\zeta^{\bar{1}}{}_r = 0, \qquad {}^cu_t = {}^cU^1\, {}^c\zeta^{\bar{1}}{}_t = -\ {}^cU^1 e^z r \tag{4.15d}$$

It is

$$ {}^cu_{x,\bar{y}} = -\ {}^cU^1 e^z y_{,\bar{y}} = -\ {}^cU^1 e^z \neq -\ {}^cu_{y,\bar{x}} = {}^cU^1 e^z x_{,\bar{x}} = {}^cU^1 e^z \tag{4.15e}$$

$$ {}^cu_{r,t} = 0 \neq -\ {}^cu_{t,r} = {}^cU^1 e^z \tag{4.15f}$$

and the motion denoted by the motive ${}^cU^1$ is contra-formal and represents a mode of deformation which implies displacements converging to the poles of the position vectors $r$ (Fig. 4.3a2).

## 4.5. Normal Force Composants and Motives

### 4.5.1. Polar Vector Composants and Motives

The quaternal decomposition of the component $f_z$ of $f$ which is normal to the plane of the cross section is shown in (Fig. 4.4). The system of forces shown in (Fig. 4.4a) corresponds to the polar vector composant, furthermore referred as normal force

$$ {}_cF_z = \sum_{n=1}^{4} {}_nf_z = f_z, \tag{4.16}$$

with lever-arm and shape function

$$ {}_c\xi_{z\bar{z}} = {}_c\zeta_{\bar{z}z} = 1 \tag{4.16a}$$

The displacements of the points for the motive ${}_cU_z$ are constant

$$ {}_cu_z = {}_cU_z\, {}_c\zeta_{\bar{z}z} = {}_cU_z = \tag{4.16b}$$

and constitute a rigid body translation along the $z$-axis.

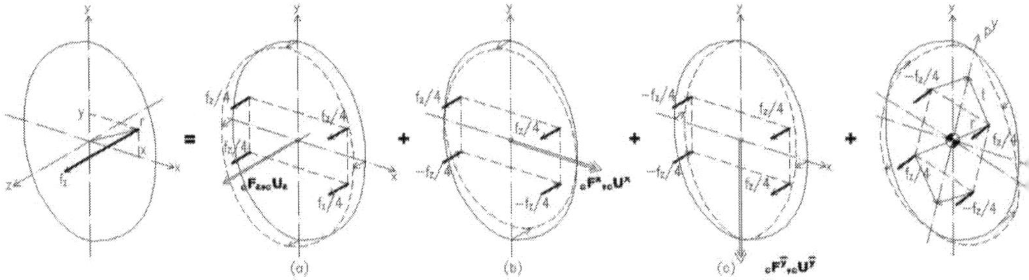

Figure 4.4 - Normal Force Composants and Motives

The above expressions correspond to the case of ${}_cF_z$ constant along the axis $z$ of the beam. For ${}_cF_z$ variable with respect to $z$ the following higher degree form of lever-arm and shape functions contributes to the representation of the normal force $F_z$

$$ {}_c\xi_{z\bar{z}} = {}_c\zeta_{\bar{z}z} = \left(-\frac{r^2}{2}\right)^1 \qquad (4.16c) $$

The displacements of the points for the corresponding motive ${}_cU_z$ are

$$ {}_cu_z = -\, {}_cU_z \frac{r^2}{2} \qquad (4.16d) $$

The relations (4.16d) is the equations of a paraboloid of revolution with axis passing from the origin of the reference system and represents the deformed shape of the cross section.
The complete expression for the lever-arm and shape function will be

$$ {}_c\xi_{z\bar{z}} = {}_c\zeta_{\bar{z}z} = a^1 - \beta^1 \frac{r^2}{2} \qquad (4.16e) $$

where $a^1$ and $\beta^1$ are real scalar coefficients.

### 4.5.2. Axial Vector Composants and Motives

For the systems of forces shown in (Figs. 4.4b, c) the polar vector composant is zero and the following axial vector composants are attributed to them, furthermore referred as *bending or flexural moments*

$$ {}_cF^x = \sum_{n=1}^{4} {}_ny\, {}_nf_z = (yf_z)^x, \qquad {}_cF^{\bar{y}} = \sum_{n=1}^{n} {}_nx\, {}_nf_z = (xf_z)^{\bar{y}} \qquad (4.17) $$

with lever-arm and shape functions

$$ {}_c\xi^x{}_{\bar{z}} = {}_c\zeta^{\bar{x}}{}_z = y, \qquad {}_c\xi^{\bar{y}}{}_{\bar{z}} = {}_c\zeta^y{}_z = x \qquad (4.17a) $$

The displacements of the points for the corresponding motives ${}_cU^x$ and ${}_cU^{\bar{y}}$ are

$$ {}^x_cu_z = {}_cU^x\, {}_c\zeta^{\bar{x}}{}_z = \left({}_cU^x y\right)_z, \qquad {}^{\bar{y}}_cu_z = {}_cU^{\bar{y}} \zeta^y{}_z = \left({}_cU\ \ x\right)_z \qquad (4.17b) $$

and constitute rigid body rotations around the axes $x$ and $y$ respectively.

The entities

$$|\,{}_cF_z| = |f_z|, \quad |\,{}_cF^A| = |\,{}_cF^x + \,{}_cF^{\bar{y}}| = |(x+y)f_z| = |rf_z| = |r \times f_z| \tag{4.18}$$

are invariant under orthogonal transformations and the motives ${}_cU_z$, ${}_cU^x$ and ${}_cU^{\bar{y}}$ represent conformal motions on the $0xy$ plane.

### 4.5.3. Imaginary Scalar Composants and Motives

For the system of forces shown in (Fig. 4.4d) no polar or axial vector as well as no real scalar composant is attributed and it is represented by an imaginary scalar composant. The scalar value must be independent of the reference system and is defined in the polar systems of the position vectors ${}_nr$. The imaginary scalar composant, furthermore referred as *bi-moment* is

$$F_e = \sum_{n=1}^{4} \left( {}_nr\, {}_nt \right) f_z = ((rt)^z f_z)_e \tag{4.19}$$

with lever-arm and shape functions

$$\xi_{e\bar{z}} = \zeta_{\bar{e}z} = rt \tag{4.19a}$$

The displacements, of any point $r$ for the motive $U_e$, furthermore refereed as *warping* are

$$u_z = U_e \zeta_{\bar{e}z} = (U_e (rt)^z)_z \tag{4.19b}$$

and represent a deformed shape of the cross section consisting of an hyperbolic paraboloid surface.

### 4.5.4. Contra-formal Normal Composants and Motives

The motives ${}_cU_z$, ${}_cU^x$, ${}_cU^{\bar{y}}$ and $F_e$ corresponding to the normal force composants ${}_cF_z$, ${}_cF^x$, ${}_cF^{\bar{y}}$ and $F_e$ respectively represent motions which are conformal on the plane $0xy$ of the cross section. The systems of forces shown in (Figs. 4.3a, b, c, d) result from the quaternal decomposition of the single force $f_z$. For that reason a further replacement of each one of them by a superposition of two systems, the one corresponding to conformal and the other to contraformal motion, similar to those resulting for the lateral force composants, is not furthermore possible. Nevertheless composants and motives corresponding to contra-formal motions are obtained from the introduction of additional forms or lever-arm and shape functions beyond the fundamental ones, which are compatible with the systems of forces shown in (Fig. 4.3). These additional forms result by using the following remark.

The lever-arm and shape functions of the conformal lateral force composants are partial derivatives of lever-arm and shape functions of conformal normal force composants according to the following relations.

From Eqs. (4.14b), (4.16c):

$$ {}_c\xi^1{}_{\bar{r}} = \left( {}_c\xi_{z\bar{z}} \right)^1{}_{\bar{r}} = \left( -\frac{r^2}{2} \right)^1{}_{,\bar{r}} = -r \qquad (4.20a)$$

From Eqs. (4.6a), (4.17a):

$$ {}_c\xi_{x\bar{x}} = \left( {}_c\xi^{\bar{y}}{}_{\bar{z}} \right)_{x,\bar{x}} = x_{,\bar{x}} = 1 \qquad (4.20b)$$

From Eqs. (4.8a), (4.17a):

$$ {}_c\xi_{y\bar{y}} = \left( {}_c\xi^{x}{}_{\bar{z}} \right)_{y,\bar{y}} = y_{,\bar{y}} = 1 \qquad (4.20c)$$

From Eqs. (4.12b), (4.19a):

$$ {}_c\xi^z{}_{\bar{t}} = \left( {}_c\xi_{e\bar{z}} \right)_{,\bar{t}} = (rt)_{,\bar{t}} = r \qquad (4.20d)$$

For the normal force composants and motives corresponding to contra-formal motions the following appropriate forms of lever-arm and shape functions are introduced which satisfy with the functions for contra-formal lateral force composants relations similar to those represented in Eqs. (4.20a, b, c, d).

For the composant

$$ {}^cF_z = {}^c\xi_{z\bar{z}} f_z \quad \text{the function} \quad {}^c\xi_{z\bar{z}} = {}^c\zeta_{\bar{z}z} = -\frac{t^2}{2} \qquad (4.21)$$

is introduced resulting to

$$({}^{c}\xi_{z\bar{z}})^{1}{}_{,\bar{t}} = \left(-\frac{t^2}{2}\right)^{1}_{,\bar{t}} = -t = {}^{c}\xi^{1}{}_{\bar{t}} \tag{4.21a}$$

For the composant $\quad {}^{c}F^{x} = {}^{c}\xi^{x}{}_{\bar{z}}f_{z}$

the function $\quad {}^{c}\xi^{x}{}_{\bar{z}} = {}^{c}\zeta^{\bar{x}}{}_{z} = y|r| = y\sqrt{-x^2-y^2} \qquad (4.22)$

is introduced resulting to

$$({}^{c}\xi^{x}{}_{\bar{z}})_{y,\bar{x}} = \left(y\sqrt{-x^2-y^2}\right)_{y,\bar{x}} = -\frac{xy}{\sqrt{-x^2-y^2}} = y\hat{x} = {}^{c}\xi_{y\bar{x}}$$
(4.22a)

For the composant
${}^{c}F^{\bar{y}} = {}^{c}\xi^{\bar{y}}{}_{\bar{z}}f_{z}$ the function ${}^{c}\xi^{\bar{y}} = {}^{c}\zeta^{y}{}_{z} = x|r| = x\sqrt{-x^2-y^2}$
(4.23)

is introduced resulting to

$$({}^{c}\xi^{\bar{y}}{}_{\bar{z}})_{x,\bar{y}} = \left(x\sqrt{-x^2-y^2}\right)_{x,\bar{y}} = -\frac{xy}{\sqrt{-x^2-y^2}} = x\hat{y} = {}^{c}\xi_{x\bar{y}} \tag{4.23a}$$

From the composant which is common for both, the conformal and the contra-formal motion

$F_e = \xi_{e\bar{z}}f_z$ with the function $\xi_{e\bar{z}} = \zeta_{\bar{e}z} = rt \qquad (4.24)$

results the lever-arm and shape function

$$({}^{c}\xi_{e\bar{z}})^{z}{}_{,\bar{r}} = (rt)^{z}{}_{,\bar{r}} = -t = {}^{c}\xi^{z}{}_{\bar{r}} = {}^{c}\zeta^{\bar{z}}{}_{r} \tag{4.24a}$$

for the lateral force composant and motive ${}^{c}F^{z}$ and ${}^{c}U^{z}$.

## 4.6. The Complete System of State Parameters

In Table 4.1 are assembled the composants and motives separately for the normal and lateral forces, as well as the corresponding to them fundamental lever-arm and shape functions. From the table becomes apparent that the composants and motives for the lateral and for the normal forces are elements of a quaternion and of a pseudo-quaternion respectively and separately for those corresponding to conformal and those corresponding *to* contraformal motions. The complete set of parameters are elements of an octonion.

**Table 4.1 - State Parameters of 2-D Particles**

| Normal Forces | | | | Lateral Forces | | | |
|---|---|---|---|---|---|---|---|
| Axial Force | ${}_cF_z$<br>${}^cF_z$ | ${}_cU_z$<br>${}^cU_z$ | ${}_c\xi_{z\bar{z}} = a^1 - \beta^1 \frac{r^2}{2}$<br>${}^c\xi_{z\bar{z}=-\frac{t^2}{2}}$ | Plane Di-force | ${}_cF^1$<br>${}^cF^1$ | ${}_eU^1$<br>${}^cU^1$ | ${}_c\xi^1{}_{\bar{r}} = \left({}_c\xi_{z\bar{z}}\right)^1{}_{,\bar{r}} = -\beta^1 r$<br>${}^c\xi^1{}_{\bar{t}} = \left({}^c\xi_{z\bar{z}}\right)^1{}_{,\bar{t}} = -t$ |
| Bending Moments | ${}_cF^x$<br>${}^cF^x$ | ${}_cU^x$<br>${}^cU^x$ | ${}_c\xi^x{}_{\bar{z}} = y$<br>${}^c\xi^x{}_{\bar{z}} = y\lvert r\rvert$ | Shear Forces | ${}_cF_y$<br>${}^cF_y$ | ${}_cU_y$<br>${}^cU_y$ | ${}_c\xi_{y\bar{y}} = \left({}_c\xi^x{}_{,\bar{z}}\right)_{y,\bar{y}} = 1$<br>${}^c\xi_{y\bar{x}} = \left({}^c\xi^x{}_{\bar{z}}\right)_{y,\bar{x}} = y\hat{x}$ |
| | ${}_cF^{\bar{y}}$<br>${}^cF^{\bar{y}}$ | ${}_cU^{\bar{y}}$<br>${}^cU^{\bar{y}}$ | ${}_c\xi^{\bar{y}}{}_{\bar{z}} = x$<br>${}^c\xi^{\bar{y}}{}_{\bar{z}} = x\lvert r\rvert$ | | ${}_cF_x$<br>${}^cF_x$ | ${}^cU_x$<br>${}^cU_x$ | ${}_c\xi_{x\bar{x}} = \left({}_c\xi^{\bar{y}}{}_{\bar{z}}\right)_{x,\bar{x}} = 1$<br>${}^c\xi_{x\bar{y}} = \left({}^c\xi^{\bar{y}}{}_{\bar{z}}\right)_{x,\bar{y}} = x\hat{y}$ |
| Bi-moment | $F_e$ | $U_e$ | $\xi_{e\bar{z}} = rt$ | Torsional-Distortional Moment | ${}_cF^z$<br>${}^cF^z$ | ${}_cU^z$<br>${}^cU^z$ | ${}_c\xi^z{}_{\bar{t}} = (\xi_{e\bar{z}})^z{}_{,\bar{t}} = r$<br>${}^c\xi^z{}_{\bar{r}} = (\xi_{e\bar{z}})^z{}_{,\bar{r}} = -t$ |

## 4.7. Loads and Stresses

On the material points of a particle the following kinds of polar vector forces are applied.

The *inertial forces*

$$dg_p = dg_x + dg_y + dg_z = g_{p\bar{e}} dV_e = a_p p_{\bar{e}} dV_e \tag{4.25}$$

where : $p_{\bar{e}}$ is the mass density of the material

$a_p = a_x + a_y + a_z$ is the acceleration induced by external potential fields

$g_{p\bar{e}} = g_{x\bar{e}} + g_{y\bar{e}} + g_{z\bar{e}}$ the inertial force per unit volume

The composants of the inertial forces applied on the totality of the material points of the particle are denoted by the symbol $G_I$, where $I$ represents one of the eight indices of the octonion elements.
The *external loads* applied on the boundary surface along the perimeter $S$

$$dq_p = dq_x + dq_y + dq_z = (q_x{}^{\bar{x}} dA^x)_x + \left(q_y{}^{\bar{y}} dA^y\right)_y + (q_z{}^{\bar{n}} dA^n)_z \quad (4.26)$$

where : $q_x{}^{\bar{x}} = q_y{}^{\bar{y}} + q_z{}^{\bar{n}}$ are the external loads per unit surface $A^x, A^y$ & $A^n$ respectively.

The composants of the external loads applied on the whole boundary surface are denoted by the symbol $Q_I$.

Two sets of stresses developed on the faces of the material points are distinguished, the interacting and the hidden. (Fig. 4.3)

In the *interacting* set the stresses are contained

$$\sigma_x{}^{\bar{z}}, \quad \sigma_y{}^{\bar{z}} \ and \ \sigma_z{}^{\bar{z}}$$

developed on the interface $A^z$ with the adjacent particle at the point $z$ of the axis of the beam and the stresses

$$\sigma_x{}^{\bar{z}} + \sigma_x{}^{\bar{z}}{}_{,\bar{z}} d_z, \qquad \sigma_y{}^{\bar{z}} + \sigma_y{}^{\bar{z}}{}_{,\bar{z}} d_z, \qquad \sigma_z{}^{\bar{z}} + \sigma_z{}^{\bar{z}}{}_{,\bar{z}} d_z$$

developed on the second interface with the adjacent particle at the point $z + dz$ of the axis.

In the *hidden set* are contained the stresses
$$\sigma_x{}^{\bar{x}}, \qquad \sigma_y{}^{\bar{x}}, \qquad \sigma_z{}^{\bar{x}}$$

developed on the face $dA^x = (dydz)^x$ of the material points at the position $x$, the stresses

$$\sigma_x{}^{\bar{x}} + \sigma_x{}^{\bar{x}}{}_{,\bar{x}}\, dx, \qquad \sigma_y{}^{\bar{x}} + \sigma_y{}^{\bar{x}}{}_{,\bar{x}}\, d\bar{x}, \qquad \sigma_z{}^{\bar{x}} + \sigma_z{}^{\bar{x}}{}_{,\bar{x}}\, dx$$

developed on the face $dA^x$ at the position $x + dx$, the stresses

$$\sigma_x{}^{\bar{y}}, \qquad \sigma_y{}^{\bar{y}}, \qquad \sigma_z{}^{\bar{y}}$$

developed on the face $dA^y = (dzdx)^y$ of the material point at the position $y$ and the stresses

$$\sigma_x{}^{\bar{y}} + \sigma_x{}^{\bar{y}}{}_{,\bar{y}}\, dy, \qquad \sigma_y{}^{\bar{y}} + \sigma_y{}^{\bar{y}}{}_{,\bar{y}}\, dy, \qquad \sigma_z{}^{\bar{y}} + \sigma_z{}^{\bar{y}}{}_{,\bar{y}}\, dy$$

developed on the face $dA^y$ at the position $y + dy$.
The polar vector resultant to the stresses developed on a material point is

$$d\sigma_p = d\sigma_x + d\sigma_y + d\sigma_z \qquad (4.27a)$$

with

$$d\sigma_i = \sum_j \left(\sigma_i{}^{\bar{J}}[r_j + dr_j] - \sigma_i{}^{\bar{J}}[r_j]\right) dA^j =$$

$$= \sum_j \left(\left(\sigma_i{}^{\bar{J}}{}_{,\bar{J}} dr_j\right) dA^j\right)_i = \left(-\sum_j \theta_{ij} \sigma_i{}^{\bar{J}}{}_{,\bar{J}}\right) dV_e \qquad (4.27b)$$

The composants of the stresses developed on the faces of the particle are denoted by the symbol $N_I$.

The equilibrium condition of the forces applied on the material points is

$$d\sigma_i + dg_i = 0 \qquad (4.28a)$$

or, after replacing from Eqs. (4.20), (4.21) and (4.22b),

$$\sum_j \sigma_i{}^{\bar{J}}{}_{,\bar{J}} + g_{i\bar{e}} = 0 \qquad (4.28b)$$

and in detailed form

$$-\sigma_x{}^{\bar{x}}{}_{,\bar{x}} + \sigma_x{}^{\bar{y}}{}_{,\bar{y}} + \sigma_x{}^{\bar{z}}{}_{,\bar{z}} + g_{x\bar{e}} = 0 \qquad (4.29a)$$

$$\sigma_y{}^{\bar{x}}{}_{,\bar{x}} - \sigma_y{}^{\bar{y}}{}_{,\bar{y}} + \sigma_y{}^{\bar{z}}{}_{,\bar{z}} + g_{y\bar{e}} = 0 \qquad (4.29b)$$

$$\sigma_z{}^{\bar{x}}{}_{,\bar{x}} + \sigma_z{}^{\bar{y}}{}_{,\bar{y}} - \sigma_z{}^{\bar{z}}{}_{,\bar{z}} + g_{z\bar{e}} = 0 \qquad (4.29c)$$

The following relations result from the equilibrium condition of moments

$$\sigma_x{}^{\bar{y}} = -\sigma_y{}^{\bar{x}}, \qquad \sigma_y{}^{\bar{z}} = -\sigma_z{}^{\bar{y}}, \qquad \sigma_z{}^{\bar{x}} = -\sigma_x{}^{\bar{z}} \qquad (4.30)$$

From the equilibrium condition of the forces applied on the material points lying on the perimeter $S$ of the cross section, after omitting the higher order infinitesimals $(\sigma_i{}^{\bar{z}}{}_{,\bar{z}})dV_e$ and $g_{i\bar{e}}dV_e$, the following boundary conditions result, relating stresses with external loads. (Fig. 4.4.)

$$(q_x{}^{\bar{x}})_e dy = (\sigma_x{}^{\bar{y}})_{\bar{z}} dx + (\sigma_x{}^{\bar{x}})_e dy \qquad (4.31a)$$

$$(q_y{}^{\bar{y}})_e dx = (\sigma_y{}^{\bar{y}})_e dx + (\sigma_y{}^{\bar{x}})_z dy \qquad (4.31b)$$

$$(q_z{}^{\bar{n}})_{\bar{s}} ds = -(\sigma_z{}^{\bar{y}})_x dx + (\sigma_z{}^{\bar{x}})_{\bar{y}} dy \qquad (4.31c)$$

From the Green's theorem in the plane, which is a special case of the Gaussian theorem, applied for the Eqs. (4.25a, b, c) the following relations result

$$\oint_S (\xi q_x{}^{\bar{x}}) dy = \oint_S \left( (\xi \sigma_x{}^{\bar{y}})_{\bar{z}} dx + (\xi \sigma_x{}^{\bar{x}})_e dy \right) =$$

$$= \iint_{A^z} \left( (\xi \sigma_x{}^{\bar{x}})_{,\bar{x}} - (\xi \sigma_x{}^{\bar{y}})_{,\bar{y}} \right) dA^z \qquad (4.32a)$$

$$\oint_S (\xi q_y{}^{\bar{x}}) dx = \oint_S \left( (\xi \sigma_y{}^{\bar{y}}) dx + (\xi \sigma_y{}^{\bar{x}}) dy \right)$$

$$= \iint_{A^z} \left( -(\xi \sigma_y{}^{\bar{x}})_{,\bar{x}} + (\xi \sigma_y{}^{\bar{y}})_{,\bar{y}} \right) dA^z \qquad (4.32b)$$

$$\oint_S (\xi q_z{}^{\bar{n}})ds = \oint_S \left(-(\xi\sigma_z{}^{\bar{y}})dx + (\xi\sigma_z{}^{\bar{x}})dy\right) =$$

$$= \iint_{A^z} \left(((\xi\sigma_z{}^{\bar{x}})_{,\bar{x}} + (\xi\sigma_z{}^{\bar{y}})_{,\bar{y}}\right)dA^z \tag{4.32c}$$

where $A^z$ is the surface of the cross section included in the perimeter *S*, $dA^z = (dxdy)^z$and $\xi = \xi[x,y]$ a leverarm function.

## 4.8. Interacting Composants

### 4.8.1. Normal Interacting Composants

The composants of the interacting stresses $\sigma_z{}^{\bar{z}}$ normal to the interface between adjacent particles, which are developed on the totality of the material points are represented by the surface integral over the total area $A^z$ of the particle:

$$N_I = \int_{A^z} \xi_{I\bar{z}}(\sigma_z{}^{\bar{z}}dA^z)_z \tag{4.33}$$

where *I* represents one of the pseudo-quaternion indices $(z, {}^{x,}\ {}^{\bar{y}}, e)$ and $\xi_{I\bar{z}}$ the leverarm functions shown in Table 4.1 for the normal force composants which are functions of the coordinates $x, y$ or $r, t$. For the above values of I the *interacting normal composants* $N_z$ (axial force), $N^{\bar{y}}, N^x$ (bending moments ) and $N_e$ (bimoment) are represented by the expression (4.33).

### 4.8.2. Lateral Interacting Composants

The composants of the interacting stresses $\sigma_x{}^{\bar{z}}$ and $\sigma_y{}^{\bar{z}}$ lateral to the axis of the beam, which are likewise developed on the totality of the material points of the particle are represented by the sum of the surface integrals

$$N_J = \int_{A^z} \xi_{J\bar{x}}((\sigma_x{}^{\bar{z}})_y dA^z)_x + \int_{A^z} \xi_{J\bar{y}}\left((\sigma_y{}^{\bar{z}})_{\bar{x}} dA^z\right)_y \tag{4.34}$$

where $J = I\bar{z}$ represents one of the quaternion indices ( $^1$, $y$, $x$, $^z$ ) and $\xi_{J\bar{x}} = \xi_{I\bar{z},\bar{x}}$, $\xi_{J\bar{y}} = \xi_{I\bar{z},\bar{y}}$ the leverarm functions shown in Table 4.1 for the lateral force composants which are functions of the coordinates $x, y$ or $r, t$. For the above values of $J$ the *interacting lateral composants* $N^1$ (diforce) $N_x, N_y$ (shear forces) and $N^z$ (tortional and distortional moments) are represented by the expression (4.34).

An important relation between the normal, the lateral interacting composants and the normal force composants of the inertia and external loads results from the following derivation procedure applied on the normal interacting composant $N_I$. It is

$$N_{I,\bar{z}} = -\int_{A^z} \xi_{I\bar{z}}((\sigma_z{}^{\bar{z}}{}_{,\bar{z}})^z dA^z) = -\int_{A^z} (\xi_{I\bar{z}}\sigma_z{}^{\bar{z}}{}_{,\bar{z}})dA^z \qquad (4.35a)$$

After replacing $\sigma_z{}^{\bar{z}}{}_{,\bar{z}}$ according to the Eq. (4.29c) the above expression (4.35a) takes the form

$$N_{I,\bar{z}} = -\int_{A^z} \xi_{I\bar{z}}(\sigma_z{}^{\bar{x}}{}_{,\bar{x}} + \sigma_z{}^{\bar{y}}{}_{,\bar{y}})\, dA^z - \int_{A^z} (\xi_{I\bar{z}} g_{z\bar{e}})dA^z \qquad (4.35b)$$

From the derivation by parts applied on the first integral results

$$N_{I,\bar{z}} = -\int_{A^z} (\xi_{I\bar{z},\bar{x}}\sigma_z{}^{\bar{x}} + \xi_{I\bar{z},\bar{y}}\sigma_z{}^{\bar{y}})dA^z - \int_{A^z} ((\xi_{I\bar{z}}\sigma_z{}^{\bar{x}})_{,\bar{x}} + (\xi_{I\bar{z}}\sigma_z{}^{\bar{y}})_{,\bar{y}})\, dA^z - G_{I\bar{z}} \qquad (4.35c)$$

The Eq. (4.35c), after taking into account the Eqs. (4.30), (4.32c) and the relations $\xi_{I\bar{z},\bar{x}} = \xi_{J\bar{x}}$, $\xi_{I\bar{z},\bar{y}} = \xi_{I\bar{y}}$, takes the form

$$N_{I,\bar{z}} = -\int_{A^z} (\xi_{J\bar{x}}\sigma_x{}^{\bar{z}} + \xi_{J\bar{y}}\sigma_y{}^{\bar{z}})\, dA^z - \oint_S ((\xi_{I\bar{z}}\sigma_z{}^{\bar{x}})dy + (\xi_{I\bar{z}}\sigma_z{}^{\bar{y}})dx) - G_{I\bar{z}}$$

$$= -\int_{A^z} \xi_{J\bar{x}}(\sigma_x{}^{\bar{z}} dA^z)_x + \int_{A^z} \xi_{J\bar{y}} \left(\sigma_y{}^{\bar{z}} dA^z\right)_y - \oint_S \xi_{I\bar{z}}(q_z{}^{\bar{n}} ds) - G_{I\bar{z}}$$

(4.35d)

According to the Eq. (4.34) the expression (4.35d) takes the final form

$$N_J = N_{I,\bar{z}} + Q_{I\bar{z}} + G_{I\bar{z}} \qquad (4.35)$$

where

$$G_{I\bar{z}} = \int_{A^z} \xi_{I\bar{z}}((g_{z\bar{e}})^{\bar{z}} dA^z)^1 \qquad (4.36)$$

is the normal composant per unit length of the beam of the inertial forces $g_{z\bar{e}}$ applied on the totality of the material points of the particle, and

$$Q_{I\bar{z}} = \oint_S \xi_{I\bar{z}}((q_z{}^{\bar{n}})_{\bar{s}} ds)^1 = \oint_S \left((\xi_{I\bar{z}}\sigma_z{}^{\bar{x}})dy - (\xi_{I\bar{z}}\sigma_z{}^{\bar{y}})dx\right) =$$

$$= \int_{A^z} \left((\xi_{I\bar{z}}\sigma_z{}^{\bar{x}})_{,\bar{x}} + (\xi_{I\bar{z}}\sigma_z{}^{\bar{y}})_{,\bar{y}}\right) dA^z \qquad (4.37)$$

is the normal composant per unit length of the beam of the external loads $q_z{}^{\bar{n}}$
.

For the different interacting composants the general expression (4.35) takes the following specific forms:

$N^1 = N_{z,\bar{z}} + Q_{z\bar{z}} + G_{z\bar{z}}$ : Di-force to axial force

$N_x = N^{\bar{y}}{}_{,\bar{z}} + Q^{\bar{y}}{}_{\bar{z}} + G^{\bar{y}}{}_{\bar{z}}$

: Shear forces to bending moments

$N_y = N^{x}{}_{,\bar{z}} + Q^{x}{}_{\bar{z}} + G^{x}{}_{z}$

$N^z = N_{e,\bar{z}} + Q_{e\bar{z}} + G_{e\bar{z}}$ :Torsional or Distortional moment to bi-moment

## 4.9. Hidden Composants

### 4.9.1. Normal Hidden Composants

In the present state of Structural Mechanics the treatment of the stress composants is limited to the six polar and axial vector interacting ones, while the scalar of them are ignored, the di-force $N^1$ totally and the bimoment $N_e$ partially. The work performed by them for the corresponding motives equals the work done by the interacting stresses $\sigma_x{}^{\bar{z}}$, $\sigma_y{}^{\bar{z}}$ and $\sigma_z{}^{\bar{z}}$ developed at the interfaces $A^z$ between adjacent particles for the corresponding strains. By this limitation the work done by the rest hidden stresses $\sigma_x{}^{\bar{x}}, \sigma_y{}^{\bar{x}}, \sigma_z{}^{\bar{x}}$ and $\sigma_x{}^{\bar{y}}, \sigma_y{}^{\bar{y}}, \sigma_z{}^{\bar{y}}$ developed at the interfaces $dA^x$ and $dA^y$ respectively between the material points in the interior of the particle is ignored although it is actually performed for the corresponding strains. The system of forces due to the hidden stresses developed, which are applied on the material points, can be represented by a set of *hidden composants*, which are distinguished in *normal* and *lateral* ones.

In the axial direction $z$ of the beam the hidden stresses $\sigma_z{}^{\bar{x}}$ and $\sigma_z{}^{\bar{y}}$ are developed. According to Eqs. (4.27b) and (4.29) the force in the same direction applied on a material point

$$dV_e = (dA^x dx)_e = (dA^y dy)_e = (dA^z dz)_e$$

lying in the position $r = x + y$, is

$$d\sigma_z = \left(\left(\left((\sigma_z{}^{\bar{x}})_{\bar{y},\bar{x}}\right)^{\bar{z}} dx\right)_{\bar{y}} dA^x\right)_z + \left(\left(\left((\sigma_z{}^{\bar{y}})_{x,\bar{y}}\right)^{\bar{z}} dy\right)_x dA^y\right)_z =$$

$$= \left(\left(\left((\sigma_z{}^{\bar{x}})_{\bar{y},\bar{x}} + (\sigma_z{}^{\bar{y}})_{x,\bar{y}}\right)^{\bar{z}} dA^z\right)^1 dz\right)_z =$$

$$= \left(\left(\left(-(\bar{\sigma}_z{}^{\bar{z}})_{e,\bar{z}} + (\bar{g}_{z\bar{e}})\right)^{\bar{z}} dA^z\right)^1 dz\right)_z \qquad (4.38)$$

The normal hidden composant $dH_I = H_{I\bar{z}} dz$, over the length $dz$ of the particle, where $I$ is one of the pseudoquaternion indices $({}_z\ ,\ {}^{\bar{y}},\ {}^{x}, e)$ and $H_{I\bar{z}}$ the expression of $H_I$ per unit length of the beam, will be:

$$dH_I = H_{I\bar{z}} dz = \int_{(A^z d_z)} \xi_{I\bar{z}} d\sigma_z =$$

$$= \int_{(A^z d_z)} \left( \xi_{I\bar{z}} \left( \left( \left( (\sigma_z{}^{\bar{x}})_{\bar{y},\bar{x}} + (\sigma_z{}^{\bar{y}})_{x,\bar{y}} \right)^{\bar{z}} dA^z \right)^1 dz \right)_z \right)_I =$$

$$= \int_{(A^z dz)} \left( \xi_{I\bar{z}} \left( \left( \left( -(\sigma_z{}^{\bar{z}})_{e,z} \right)^{\bar{z}} + (\bar{g}_{z\bar{e}})^{\bar{z}} \right) dA^z \right)^1 dz \right)_z \right)_I \qquad (4.39)$$

Except the normal hidden composant $dH_I$ on the particle the following additional normal composants are applied.

The resultant of the normal interacting composants applied on the two interfaces of the particle

$$\Delta N_I = N_I[z + d_z] - N_I[z] = N_{I,\bar{z}} dz$$

The normal composants

$$dG_I = G_{I\bar{z}} dz, \qquad dQ_I = Q_{I\bar{z}} dz$$

of the inertial and of the external loads respectively, where $G_{I\bar{z}}$ and $Q_{I\bar{z}}$ are their expressions per unit length of the beam already defined by the Eqs. (4.36) and (4.37).
From the equilibrium condition of the forces defined by the Eqs. (4.39a, b, c) and from Eq. (4.35) results

$$dH_I + \Delta N_I + dG_I + dQ_I = \left( H_{I\bar{z}} + N_{I,\bar{z}} + G_{I\bar{z}} + Q_{I\bar{z}} \right) dz = 0$$

or

$$dH_I = (H_{I\bar{z}} dz)_K = \left( (\bar{N}_{I,\bar{z}} + \bar{G}_{I\bar{z}} + \bar{Q}_{I\bar{z}}) dz \right)_I = \left( \bar{N}_J dz \right)_I \qquad (4.40)$$

where $N_J$ is the lateral interacting composant corresponding to $J = I\bar{z}$ and $K = 1$.

The analytical expression for $H_{I\bar{z}}$ resulting from the derivation procedure is

$$H_{I\bar{z}} = dH_I\widehat{dz} = \int_{A^z} \left(\xi_{I\bar{z},\bar{x}}(\sigma_z{}^{\bar{x}})_{\bar{y}} + \xi_{I\bar{z},\bar{y}}(\sigma_z{}^{\bar{y}})_x\right)dA^z =$$

$$= \int_{A^z} \left( (\xi_{I\bar{z},\bar{x}}dx)((\sigma_z{}^{\bar{x}})_y dA^x)_z + (\xi_{I\bar{z},\bar{y}}dy)((\sigma_z{}^{\bar{y}})_x dA^y)_z \right)\widehat{dz}$$

(4.40a)

with lever-arm functions of $dH_I$ the terms of the total differential

$$d\xi_{I\bar{z}} = \xi_{I\bar{z},\bar{x}}dx + \xi_{I\bar{z},\bar{y}}dy \tag{4.40b}$$

### 4.9.2. Lateral Hidden Composants

In the lateral directions $x$ and $y$ of the beam the hidden stresses $\sigma_x{}^{\bar{x}}$, $\sigma_x{}^{\bar{y}}$ and $\sigma_y{}^{\bar{y}}$, $\sigma_y{}^{\bar{x}} = -\sigma_x{}^{\bar{y}}$ respectively are developed. According to Eqs. (4.27b) and (4.29) the forces in the same directions applied on a material point are

$$d\sigma_x \left( \left( \left( (\sigma_x{}^{\bar{x}})_{e,\bar{x}} \right)^x dx \right)_e dA^x \right)_x + \left( \left( \left( (\sigma_x{}^{\bar{y}})_{\bar{z},\bar{y}} \right)^{\bar{x}} dy \right)_{\bar{z}} dA^y \right)_x =$$

$$= \left( \left( \left( (\sigma_x{}^{\bar{x}})_{e,\bar{x}} - (\sigma_x{}^{\bar{y}})_{\bar{z},\bar{y}} \right)^x dA^z \right)^y dz \right)_x =$$

$$= \left( \left( \left( -(\bar{\sigma}_x{}^{\bar{z}})_{y,\bar{z}} - \bar{g}_{x\bar{e}} \right)^x dA^z \right)^y dz \right)_x \tag{4.41a}$$

$$d\sigma_y = \left( \left( \left( (\sigma_y{}^{\bar{y}})_{e,\bar{y}} \right)^y dy \right)_e dA^y \right)_y + \left( \left( \left( (\sigma_y{}^{\bar{x}})_{z,\bar{x}} \right)^{\bar{y}} dx \right)_z dA^x \right)_y =$$

$$= \left( \left( \left( (\sigma_y{}^{\bar{y}})_{e,\bar{y}} - (\sigma_y{}^{\bar{x}})_{\bar{z},\bar{x}} \right)^y dA^z \right)^x dz \right)_y =$$

$$= \left( \left( \left( -(\bar{\sigma}_y{}^{\bar{z}})_{\bar{x},\bar{z}} - \bar{g}_{y\bar{e}} \right)^y dA^z \right)^{\bar{x}} dz \right)_y \tag{4.41b}$$

The lateral hidden composant $dH_J = H_{J\bar{z}}dz$, where $J$ is one of the quaternion indices ( $^1$, $_x$, $_y$, $^z$ ) and $H_{J\bar{z}}$ the expression of $H_J$ per unit length of the beam, will be

$$dH_J = H_{J\bar{z}}dz = \int_{(A^z dz)} \left(\xi_{J\bar{x}}d\sigma_x + \xi_{J\bar{y}}d\sigma_y\right) =$$

$$= \int_{(A^z dz)} \left( \xi_{J\bar{x}} \left( \left( \left( -\left(\bar{\sigma}_x{}^{\bar{z}}\right)_{y,\bar{z}} \right)^x dA^z \right)^y dz \right)_x \right.$$
$$+ \xi_{J\bar{y}} \left( \left( \left( -\left(\bar{\sigma}_y{}^{\bar{z}}\right)_{\bar{x},\bar{z}} \right)^y dA^z \right)^{\bar{x}} dz \right)_y$$
$$\left. + \xi_{J\bar{x}}(((-\bar{g}_{x\bar{e}})^x dA^z)^y dz)_x + \xi_{J\bar{y}} \left( \left( (-\bar{g}_{y\bar{e}})^y dA^z \right)^{\bar{x}} dz \right)_y \right)_J$$

Except the lateral hidden composant $dH_J$ on the particle the following additional lateral composants are applied:

The resultant of the lateral interacting composants applied on the two interfaces of the particle

$$\Delta N_J = N_J[z + dz] - N_J[z] = N_{J,\bar{z}} dz$$

The lateral composants

$$dG_J = G_{J\bar{z}} dz, \qquad dQ_J = Q_{J\bar{z}} dz$$

of the internal and of the external loads respectively, where $G_{J\bar{z}}$ and $Q_{J\bar{z}}$ are their expressions per unit length of the beam.

From the equilibrium condition of these forces results

$$dH_J + \Delta_{NJ} + dG_J + dQ_J = \left(H_{J\bar{z}} + N_{J,\bar{z}} + G_{J\bar{z}} + Q_{J\bar{z}}\right) dz = 0$$

or

$$dH_J = \left(H_{Jz} dz\right)_L = \left(\left(\bar{N}_{J,\bar{z}} + \bar{G}_{Jz} + \bar{Q}_{J\bar{z}}\right) dz\right)_J \qquad (4.43)$$

Analytical expression for the terms contained in Eq. (4.43) results from a derivation procedure similar to that applied for $N_{I,\bar{z}}$ according to Eqs. (4.35a, b, c). The non-associativity – alternativity property of octonions implies that the sign of the terms in the resulting sums depends on the specific expressions of $\xi_{J\bar{x}}$ and $\xi_{J\bar{y}}$ for the different $J$ indices. The extensive and complicated specific expressions for $H_{J\bar{z}}, G_{J\bar{z}}, Q_{J\bar{z}}, \xi_{L\bar{x}}, \xi_{L\bar{y}}$, which result from the same procedure are

$$H_{Jz} = dH_J\widehat{dz} = \int_{A^z} \Big(\theta_J{}^z\xi_{J\bar{x},\bar{x}}(\sigma_x{}^{\bar{x}})_e + \xi_{J\bar{x},\bar{y}}(\sigma_x{}^{\bar{y}})_{\bar{z}} + \xi_{J\bar{y},\bar{x}}\big(\sigma_y{}^{\bar{x}}\big)_z + \theta_J{}^z\xi_{J\bar{y},\bar{y}}\big(\sigma_y{}^{\bar{y}}\big)_e\Big)dA^z =$$

$$= -\left(\int_{A^z} \Big(\big((\xi_{J\bar{x},\bar{x}}dx)((\sigma_x{}^{\bar{x}})_e dA^x)_x\big) + (\xi_{J\bar{x},\bar{y}}dy)((\sigma_x{}^{\bar{y}})_{\bar{z}}dA^y)_x + \big(\xi_{J\bar{y},\bar{x}dx}\big)\Big(\big(\sigma_y{}^{\bar{x}}\big)_z dA^x\Big)_y + \big(\xi_{J\bar{y},\bar{y}}dy\big)\Big(\big(\sigma_y{}^{\bar{y}}\big)_e dA^y\Big)_y\right)_J \widehat{dz} \qquad (4.44a)$$

with lever-arm functions of $dN_L$ the terms of the total differentials

$$d\xi_{J\bar{x}} = \xi_{J\bar{x},\bar{x}}dx + \xi_{J\bar{y},\bar{y}}dy \qquad (4.44b)$$

$$d\xi_{J\bar{y}} = \xi_{J\bar{y},\bar{x}}dx + \xi_{J\bar{y},\bar{y}}dy \qquad (4.44c)$$

$$G_{J\bar{z}} = dG_J\widehat{dz} = \int_{A^z} \big(\xi_{J\bar{x}}(g_{x\bar{e}}dV_e)_x + \xi_{J\bar{y}}\big(g_{y\bar{e}}dV_e\big)\big)\widehat{dz} \qquad (4.44d)$$

$$Q_{J\bar{z}} = dQ_J\widehat{dz} = \oint_S \Big(\xi_{J\bar{x}}(q_x{}^{\bar{x}}dA^x)_x + \xi_{J\bar{y}}\big(q_y{}^{\bar{y}}dA^y\big)_y\Big)_J \widehat{dz} =$$

$$= -\int_{A^z} \Big(\theta_{(Jx)}{}^z\big(\xi_{J\bar{x}}\,(\sigma_x{}^{\bar{x}})_e\big)_{,x} - \big(\xi_{J\bar{x}}(\sigma_x{}^{\bar{y}})_{\bar{z}}\big)_{,\bar{y}} + \theta_{(Iy)}{}^z\big(\xi_{J\bar{y}}\,(\sigma_y{}^{\bar{y}})_e\big)_{,\bar{y}} - \big(\xi_{J\bar{y}}(\sigma_y{}^{\bar{x}})_{\bar{z}}\big)_{,\bar{x}}\Big)dA^z \qquad (4.44e)$$

### 4.9.3. 4.9.3 Universal Strains and Stresses

The variation of the work performed by the composants for the corresponding motives along the length $dz$ of a particle is expressed by the following relations.

For the normal and lateral interacting composants and motives:

$$\begin{aligned}\Delta_I^N W^{\bar{1}} &= N_I[z+dz]\cdot U_I[z+dz] - N_I[z]\cdot U_I[z] \\ &= \left(N_I + N_{I,\bar{z}}dz\right)\cdot\left(U_I + U_{I,\bar{z}}dz\right) - N_I\cdot U_I \\ &= N_I\cdot\left(U_{I,\bar{z}}dz\right)_I + U_I\cdot\left(N_{I,\bar{z}}dz\right)_I \end{aligned} \qquad (4.45a)$$

and in a similar manner

$$\Delta_J^N W^1 = N_J\cdot\left(U_{J,\bar{z}}dz\right)_J + U_J\left(N_{J,\bar{z}}dz\right)_J \qquad (4.45b)$$

The work performed by the normal and lateral hidden composants $dH_I$ and $dH_J$ for the same motives $U_I$ and $U_J$ respectively, after taking into account the Eqs. (4.40) and (4.43), will be

$$\Delta_I^H W^{\bar{1}} = dH_I\cdot U_I = -U_I\left(\left(N_{I,\bar{z}} + G_{I\bar{z}} + Q_{I\bar{z}}\right)dz\right)_I \qquad (4.45c)$$

$$\Delta_J^H W^{\bar{1}} = dH_J\cdot U_J = -U_J\left(\left(N_{J,\bar{z}} + G_{J\bar{z}} + Q_{J\bar{z}}\right)dz\right)_J \qquad (4.45d)$$

The work performed by the inertial and external loads composants is

$$\begin{aligned}\Delta_{(G+Q)} W^{\bar{1}} &= \left(dG_I + dQ_I\right)\cdot U_I + \left(dG_J + dQ_J\right)\cdot U_J \\ &= U_I\cdot\left(\left(G_{I\bar{z}} + Q_{I\bar{z}}\right)dz\right)_I + U_J\cdot\left(\left(G_{J\bar{z}} + Q_{J\bar{z}}\right)dz\right)_J \end{aligned} \qquad (4.45e)$$

The variation of the sum of the work performed by the interacting and the hidden composants have to be equal with that performed by the inertial and external loads, namely,

$$\Delta_I^N W^{\bar{1}} + \Delta_J^N W^{\bar{1}} + \Delta_I^H W^{\bar{1}} + \Delta_J^H W^1 = \Delta_{(G+Q)} W^{\bar{1}} \qquad (4.46a)$$

and after replacing Eqs. (4.45a to e) to Eq. (4.46a),

$$\frac{1}{2}\left(N_I \cdot (U_{I,\bar{z}}dz)_{,I} + N_J \cdot (U_{J,\bar{z}}dz)_{,J}\right) =$$
$$= U_I \cdot ((G_{I\bar{z}} + Q_{I\bar{z}})dz)_{,I} + U_J \cdot \left((G_{J\bar{z}} + Q_{J\bar{z}})dz\right) \qquad (4.46b)$$

From Eq. (4.46b) results that to the work performed by the inertial and external loads corresponds the part of the work performed by the interacting composants which is represented by the first terms $N_I \cdot (U_{I,\bar{z}}dz)$ and $N_J \cdot (U_{J,\bar{z}}dz)$ of the right side of Eqs. (4.45a, b). The factor ½ is due to the fact that the composants $N_I$ and $N_J$ are developed simultaneously with the deformations $U_{I,\bar{z}}$ and $U_{J,\bar{z}}$ respectively under a relation considered as linear along the infinitecimal length.

The omitted second terms $U_I \cdot (N_{I,\bar{z}}dz)$ and $U_J \cdot (N_{J,\bar{z}}dz)$ at the right side of Eqs. (4.45a, b), which are equal and opposite to the first terms at the right side of Eqs. (4.45c, d), represent energy quantities exchanged between the interacting and the hidden composants.

The entities $U_{I,\bar{z}}$ and $U_{J,\bar{z}}$ represent the rate of variation of the motives $U_I$ and $U_J$ along the axis $z$ of the beam and will be referred as *universal strains*. For the different values of the indices $I$ and $J$ they are defined in Structural Mechanics by the following terms and notations:

| | |
|---|---|
| $U_{z,\bar{z}}$: | Mean axial strain $\varepsilon_{zzo}$ |
| $U^{x}{}_{,\bar{z}}$ and $U^{\bar{y}}{}_{,\bar{z}}$: | Flexural curvatures $\kappa_z$ and $\kappa_y$ |
| $U_{e,\bar{z}}$: | NA |
| $U^{1}{}_{,\bar{z}}$: | NA |
| $U_{x,\bar{z}}$ and $U_{y,\bar{z}}$: | Mean lateral shear strains $\varepsilon_{xzo}$ and $\varepsilon_{yzo}$ |
| $U^{z}{}_{,\bar{z}}$: | Torsional and distortional curvature $\kappa_z$ |

The entities $N_{I,\bar{z}}$ and $N_{J,\bar{z}}$ represent the rate of variation of the composants $N_I$ and $N_J$ along the axis of the beam and will be referred as *universal stresses*. A definition of them is not available in the present state of Structural Mechanics because the hidden composants and the associated with them hidden (or dark) exchange energy are ignored.

# 5. TWO DIMENSIONAL AGGREGATES OF ONE DIMENSIONAL PARTICLES

## 5.1. Surface Systems

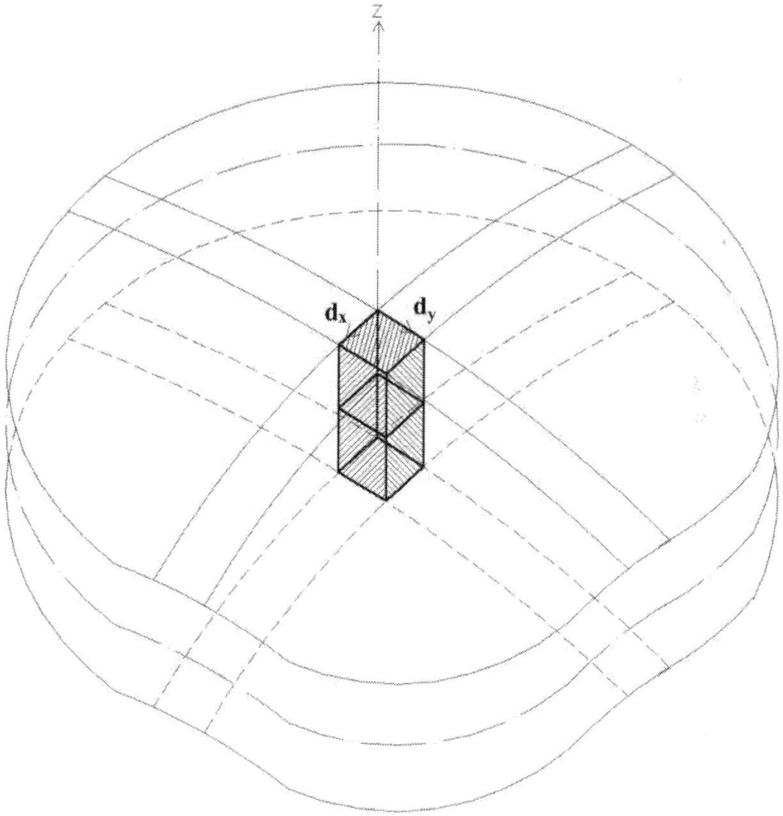


*Figure 5.1 - Particles of Plates*

As surface systems are considered are considered continuum segments having the form of a layer bounded by two quasi-parallel surfaces at small distance compared to their extent. The intermediate surface lying at equal distances from the two bounding ones is the middle surface of the layer. The position of the points on the middle surface is defined in a system of orthogonal curvilinear coordinates consisting of the geodesic lines passing through its points. The distance between the two bounding surfaces is the direction normal to the middle surface is the thickness of the layer at the corresponding point. Examples of surface systems are solid members of structures having the form of membranes, shells, plates, walls, as well as layers of fluids flowing over a spillway or ejected from a linear orifice etc.

The surface system is considered as a two dimensional aggregate of particles consisting of a linear sequence of material points $dV_e = ((dxsy)^z dz)_e$ along the thickness at any point of the middle surface (Fig. 5.1). The composants and the motives of each one particle are the system parameters of the two dimensional aggregate at the corresponding point of the middle surface. The continuity condition of the middle surface requires equal displacements, velocities, strains and strain rates to be developed at the common points on the four interfaces with the adjacent particles. This condition implies that equal stresses and impulses are likewise are likewise developed at the two interfaces, which constitute the interacting forces and momentums between the particles. As the target of the present treatise is restricted to the definition of the composants and motives, which constitute the state parameters for this specific type of particles, the investigation is simplified by considering the static case of the solid deformable surface structures. For that case the laminar state of the continuum is considered where only polar vector forces are applied and polar vector displacements are developed at the material points without the effect of momentums and velocities. The definitions for the composants and motives concerning momentums and velocities respectively, as for example for the case of flowing layers of fluids, are similar to those resulting for forces and displacements. These definitions, already established for the two dimensional particles of the beams, will be used modified according for the one dimensional particles of the surface structures with infinitesimal instead of finite width at the interfaces.

Generally the geodesic lines of the middle surface are continuous space curves and the thickness is likewise continuously variable. The analytical treatment of the state for this general form is accessible in cases where the form of the surface and the variation of the thickness are defined by normal analytical expressions. Nevertheless by using the recently developed Finite Element Method the treatment of the state of surface structures of any shape becomes accessible. The middle surfaces is replaced by a polyhedral surface consisting of triangular or rectangular plane segments which is inscribed in the middle surface. The thickness of each one plane segment is constant and equal to the average thickness of the corresponding curved segment. The desired accuracy is obtained by choosing the appropriate density of the nodes on the middle surface. Under this view the further investigation can be limited to the simpler case of plane structures with constant thickness without loss of the main target which consists to the definition of the state parameters.

## 5.2. Reference System

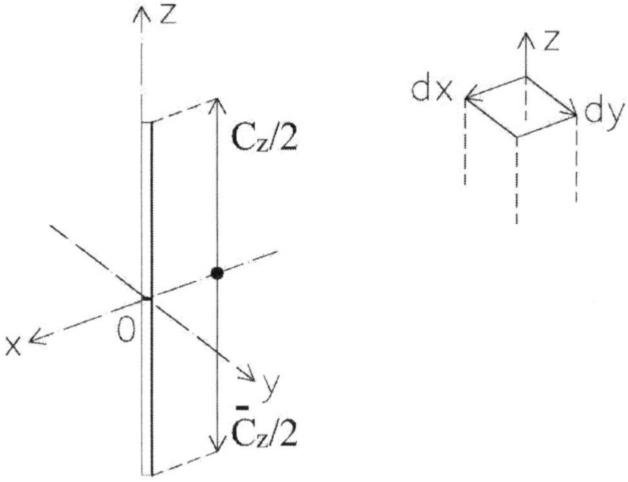


*Figure 5.2 - Reference System*

A global orthogonal reference system $oxyz$ is associated with each one linear particle, having the origin at the intersection with the middle surface, the $x$ and $y - axes$ lying on this surface and the $z - axis$ along the particle in the direction of the thickness of the surface system. The position of the material points of the particle is defined by the single coordinate $r = z$, implying for the rest coordinates $x = y = 0$. For that reason the polar systems which have been introduced for the material points belonging to three of two - dimensional particles are meaningless for the material points belonging to one – dimensional particles.

Furthermore the normal $n$ on the outer faces of the particle lying on the bounding surfaces coincides with the $z - axis$ and their infinitecimal area is

$$dA^n = (dxsy)^z = dA^z$$

## 5.3. The Tri-Dual Decomposition

The definition of the kinds of composants and motives for the one-dimensional particles is obtained through the introduction of the following tri-dual decomposition procedure, which constitutes an appropriate extension of the octal and bi-quaternal decomposition procedures introduced in the chapters 3 and 4 for the three and the two-dimensional particles respectively. For that purpose the effect of a single force $f = f_x + f_y + f_z$ on the composants is considered which is applied on a single material point lying in the position $r = z$ of the particle. Three separate dual decomposistion procedures are applied, each one concerning one of the components $f_x, f_y$ and $f_z$ of $f$ separately (Fig 5.3). At these three separate

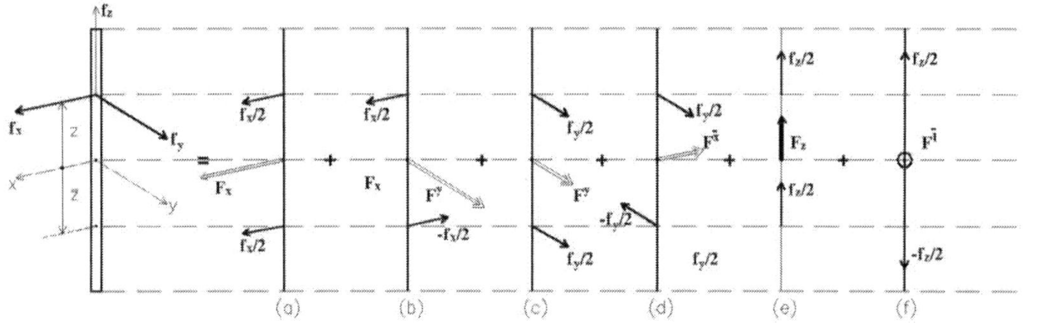


*Figure 5.3 - The Tri-Dual Composition*

procedures the corresponding component $f_i$ is replaced by the superposition of two systems each one consisting of two forces which are applied at the two points $z$ and $-z$ of the particle. At the point $z$ in both systems the component $f_i/2$ is applied while at the point $-z$ in the one system the component $f_i/2$ is applied and in the other the component $-f_i/2$. From the superposition of six systems zero components at the point $-z$ and the initial components $f_i$ and the point $z$ are obtained. The above tri-dual decomposition procedure for the definition of the composants is unique.

From the (Fig. 5.3) the following six composants and motives and the corresponding to them fundamental forms of lever-arm and shape functions are deduced:

$$\text{(Fig. 5.3a)} \qquad F_x = \xi_{x\bar{x}} f_x, \qquad \xi_{x\bar{x}} = \zeta_{\bar{x}x} = 1 \qquad (5.1a)$$

(Fig. 5.3b) $F^y = \xi^y{}_{\bar{x}}\, f_x,$ $\xi^y{}_{\bar{x}} = \zeta^{\bar{y}}{}_x = z$ (5.1b)

(Fig. 5.3c) $F_y = \xi_{y\bar{y}} f_y,$ $\xi_{y\bar{y}} = \zeta_{\bar{y}x} = 1$ (5.1c)

(Fig. 5.3d) $F^{\bar{x}} = \xi^{\bar{x}}{}_y\, f_y,$ $\xi^{\bar{x}}{}_y = \zeta^x{}_y = z$ (5.1d)

(Fig. 5.3e) $F_z = \xi_{z\bar{z}} f_z,$ $\xi_{z\bar{z}} = \zeta_{\bar{z}z} = 1$ (5.1e)

(Fig. 5.3f) $F^1 = \xi^{\bar{1}}{}_{\bar{z}}\, f_z,$ $\xi^{\bar{1}}{}_{\bar{z}} \zeta^1{}_z = z$ (5.1f)

The composants $F^z$ and $F_e$ are not included in the above system because the corresponding to them lever-arm and shape functions, which depend on the zero coordinates $x$ and $y$ of the material points, have zero values.

The above six composants correspond to conformal motions. The absence of polar system implies that contra-formal motions are not possible for the one-dimensional particles.

### 5.4. Loads and Stresses

On the material points of the one-dimensional particles the same kinds of loads with those considered in 4.6 for the plane particles of the beam are applied consisting of inertial and external polar vector forces. The inertial forces are those expressed by the relation (4.25)
Their polar vector forces $dg_i$, which are applied on each one material point $dV_e$ of the particle $(i = x, y, z)$ are

$$dg_i = g_{i\bar{e}} dV_e = -\theta_{iz}(g_{i\bar{e}} dz) dA^z = (dg_i{}^{\bar{z}}) dA^z \qquad (5.2a)$$

The axial vector and real scalar composants $dg^I$ of these forces, where $I = zi, =^{y,\bar{x},\bar{1}}$, will be

$$dg^I = dg^{(zi)} = z dg_i = \big(-z(g_{i\bar{e}}\, dz)\big) dA^z = (dg^{I\bar{z}}) dA^z \qquad (5.2b)$$

In the above expressions the symbols $dg_i{}^{\bar{z}}$ and $dg^{I\bar{z}}$ represent the values per unit area $A^z$ of $dg_i$ and $dg^I$ respectively.

The external forces are expressed by the relations

$$dq_x = (q_x{}^{\bar{z}} dA^z)_x, \qquad dq_y = \big(q_y{}^{\bar{z}} dA^z\big)_y, \qquad dq_z = (q_z{}^{\bar{z}} dA^z)_z \qquad (5.3)$$

the faces $(dxdy)^z = dA^z$ the outer material points lying on the upper and lower bounding surfaces of the system. Left indices on the symbols $q_i{}^{\bar{z}}$ , an upper on $u$ and a lower one $l$, are introduced for the distinction between the external loads ${}^u q_i{}^{\bar{z}}$ and ${}_l q_i{}^{\bar{z}}$ applied on the upper and the lower outer faces respectively of the particle.

The stresses developed on the faces of the material points are distinguished in interacting and hidden. Interacting are those developed on the faces $dA^x = (dydz)^x$ and $dA^y = (dzdx)^y$ lining on the interfaces with the adjacent particles. On the two opposite faces $dA^x$are developed the stresses

$$\sigma_x{}^{\bar{x}}, \sigma_y{}^{\bar{x}}, \sigma_z{}^{\bar{x}} \text{ and } \sigma_x{}^{\bar{x}}, \sigma_x{}^{\bar{x}}{}_{,\bar{x}} d_x, \qquad \sigma_y{}^{\bar{x}}, \sigma_y{}^{\bar{x}}{}_{,\bar{x}} d_x, \qquad \sigma_z{}^{\bar{x}}, \sigma_z{}^{\bar{x}}{}_{,\bar{x}} d_x$$

respectively, while on the two opposite faces $dA^y$ are developed the stresses

$$\sigma_x{}^{\bar{y}}, \sigma_y{}^{\bar{y}}, \sigma_z{}^{\bar{y}} \text{ and } \sigma_x{}^{\bar{y}} + \sigma_x{}^{y}{}_{,\bar{y}} dy, \qquad \sigma_y{}^{\bar{y}} + \sigma_y{}^{y}{}_{,\bar{y}} dy, \qquad \sigma_z{}^{y} + \sigma_z{}^{\bar{y}}{}_{,\bar{y}} dy$$

Hidden are the stresses

$$\sigma_x{}^{\bar{z}}, \sigma_y{}^{\bar{z}}, \sigma_z{}^{\bar{z}} \text{ and } \sigma_x{}^{\bar{z}} + \sigma_x{}^{\bar{z}}{}_{,\bar{z}} dz, \qquad \sigma_y{}^{\bar{z}} + \sigma_y{}^{\bar{z}}{}_{,\bar{z}} dz, \qquad \sigma_z{}^{\bar{z}} + \sigma_z{}^{\bar{z}}{}_{,\bar{z}} dz$$

developed on the two opposite interfaces $dA^z = (dxdy)^z$ between successive material points along the particle. The stresses satisfy the equilibrium conditions (4.29) and (4.30). The equilibrium conditions of the forces applied on the upper and the lower material points of the particle which are denoted by the upper and lower indices ${}^u$and ${}_l$ respectively result to the following boundary conditions (Fig. 5.4)

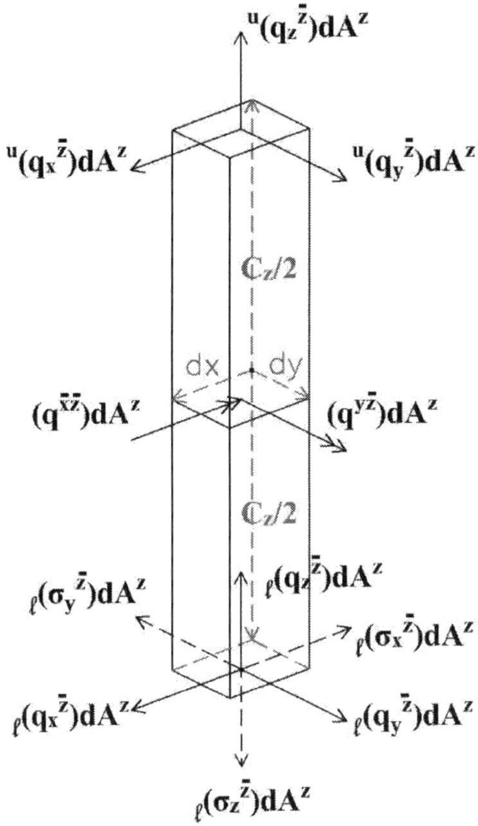


*Figure 5.4 - Loads and Stresses*

$${}^{u}q_x^{\bar z} = {}^{u}\sigma_x^{\bar z}, \qquad {}^{u}q_y^{\bar z} = {}^{u}\sigma_y^{\bar z}, \qquad {}^{u}q_z^{\bar z} = {}^{u}\sigma_z^{\bar z} \tag{5.4a}$$

$${}_{\iota}q_x^{\bar z} = {}_{\iota}\bar\sigma_x^{\bar z}, \qquad {}_{\iota}q_y^{\bar z} = {}_{\iota}\bar\sigma_y^{\bar z}, \qquad {}_{\iota}q_z^{\bar z} = {}_{\iota}\bar\sigma_z^{\bar z} \tag{5.4b}$$

and

$$q_x^{\bar z} = {}^{u}q_x^{\bar z} + {}_{\iota}q_x^{\bar z}, \qquad q_y^{\bar z} = {}^{u}q_y^{\bar z} + {}_{\iota}q_y^{\bar z}, \qquad q_z^{\bar z} = {}^{u}q_z^{\bar z} + {}_{\iota}q_z^{\bar z} \tag{5.4c}$$

$$(q^{\bar x\bar y})^y = -\left(\frac{c_z}{2}\left({}^{u}q_y^{\bar z}\right)_{\bar x} + \frac{\bar c_z}{2}\left({}_{\iota}q_y^{\bar z}\right)_{\bar x}\right)^{\bar x} = \left(-\frac{c_z}{2}\left({}^{u}\sigma_y^{\bar z} + {}_{\iota}\sigma_y^{\bar z}\right)_{\bar x}\right)^y \tag{5.4d}$$

$$(q^{y\bar z})^x = -\left(\frac{c_z}{2}\left({}^{u}q_x^{\bar z}\right)_y + \frac{\bar c_z}{2}\left({}_{\iota}q_x^{\bar z}\right)_y\right)^{\bar x} = \left(-\frac{c_z}{2}\left({}^{u}\sigma_x^{\bar z} + {}_{\iota}\sigma_x^{\bar z}\right)_y\right)^x \tag{5.4e}$$

$$\left(q^{\bar 1\bar z}\right)^z = \left(\frac{c_z}{2}\left({}^{u}q_z^{\bar z}\right)_e + \frac{\bar c_z}{2}\left({}_{\iota}q_z^{\bar z}\right)_e\right)^z = \left(\frac{c_z}{2}\left({}^{u}\sigma_z^{\bar z} + {}_{\iota}\sigma_z^{\bar z}\right)_e\right)^z \tag{5.4f}$$

### 5.5. Interacting Composants

As interacting are considered the composants of the interacting stresses developed at the interfaces with the adjacent particles. The one dimensional particles are surrounded by interfaces in the two directions $x$ and $y$, both of infinitecimal width $d_y, d_x$ and areas $dA^x = (dy)c_z, dA^y = c_z(dx)$ respectively. The six composants of the interacting stresses, which correspond to each one of the interfaces $dA^x$ and $dA^y$, have the general form

$$^x dN_I = n_{I\bar{y}} dy, \quad ^y dN_I = n_{I\bar{x}} dx \tag{5.5}$$

where $I$ represents one of the octonion indices, the upper left index denotes the interface $dA^x$ or $dA^y$ where the composant is considered and $n_{I\bar{y}}$, $n_{I\bar{x}}$ are the finite values of the composants per unit width along the axes $y$ and $x$ respectively.
The elementary resultants of the interacting stresses are divided in the following three classes: (Figs. 5.4a, b, c).

On interface $dA^x$ On interface $dA^y$

I. Normal : $^x d\sigma_x = ((\sigma_x{}^{\bar{x}})_e (dydz)^x)_x$, $^y d\sigma_y = \left((\sigma_y{}^{\bar{y}})_e (dzdx)^y\right)_y$ (5.6a)

II. Lateral : $^x d\sigma_z = \left((\sigma_z{}^{\bar{x}})_{\bar{y}} (dydz)^x\right)_z$, $^y d\sigma_z = ((\sigma_z{}^{\bar{y}})_x (dzdx)^y)_z$ (5.6c)

III. Tangential: $^x d\sigma_y = \left((\sigma_y{}^{\bar{x}})_z (dydz)^x\right)_y$, $^y d\sigma_x = ((\sigma_x{}^{\bar{y}})_{\bar{z}} (dzdx)^y)_x$ (5.6c)

According the corresponding to them composants are likewise distributed in similar classes. On the interface $dA^x$ they have the following expressions for the fundamental forms of the hebelarm functions. (The symbol $\int_c$ symbol $\int_{\bar{c}_z/2}^{c_z/2}$)

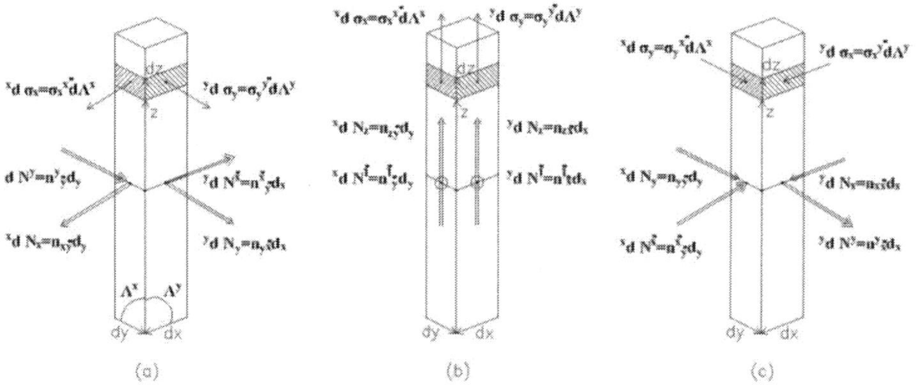


*Figure 5.5 - Interacting Composants*

*I. Normal interacting composants: (Fig. 5.5a)*

The axial force for $\xi_{x\bar{x}} = 1$

$$ {}^{x}d_{N_x} = \left(n_{x\bar{y}}\right)^{\bar{z}} dy = \int_{C} {}^{x}d\sigma_x = \int_{C} (\sigma_x{}^{\bar{x}})_e (dydz)^x = \left( \int_{C} (\sigma_x{}^{\bar{x}})_e dz \right)^{\bar{z}} dy \tag{5.7a} $$

and the bending moment for

$$ \xi^{y}{}_{\bar{x}} = z\,{}^{x}dN^{y} = \left(n^{y}{}_{\bar{y}}\right)_{\bar{e}} dy = \int_{C} z\,{}^{x}d\sigma_x = \int_{C} z((\sigma_x{}^{\bar{x}})_e (dydz)^x)_x = $$

$$ = \left( \int_{C} (-z(\sigma_x{}^{\bar{x}})_e)^{\bar{z}}\, dz \right)_e dy \tag{5.7b} $$

*II. Lateral interacting composants: (Fig. 5.5b)*

The lateral shear force for $\xi_{z\bar{z}} = 1$

$$ {}^{x}dN_z = \left(n_{z\bar{y}}\right)^{x} dy = $$

$$ = \int_{C} {}^{x}d\sigma_z = \int_{C} \left((\sigma_z{}^{\bar{x}})_{\bar{y}} (dydz)^x\right)_z = \int_{C} \left(-(\sigma_z{}^{\bar{x}})_{\bar{y}} dz\right)^{x} dy \tag{5.7c} $$

and the linear di-force for

$$\xi^{\bar{1}}{}_{\bar{z}} = z\,{}^{x}dN^{\bar{1}} = \left(n^{\bar{1}}{}_{\bar{y}}\right)_{y} dy = \int_{c} z\,{}^{x}d\sigma_{z} = \int_{c} \left(z\left((\sigma_{z}{}^{\bar{x}})_{\bar{y}}(dydz)^{x}\right)_{z}\right)^{\bar{1}} =$$

$$= \left(\int_{c} \left(-z(\sigma_{z}{}^{\bar{x}})_{\bar{y}}\right)^{\bar{x}} dz\right)_{y} dy \qquad (5.7d)$$

*III. Tangential interacting composants (Fig. 5.5c)*

The tangential shear force for $\xi_{y\bar{y}} = 1$

$${}^{x}dN_{y} = \left(n_{y\bar{y}}\right)\ dy = \int_{c} {}^{x}d\sigma_{y} = \int_{c} \left(\sigma_{y}{}^{\bar{x}}\right)_{z} (dydz)^{x} =$$

$$= \left(\int_{c} \left(-\sigma_{y}{}^{\bar{x}}\right)_{\bar{z}} dz\right)^{1} dy \qquad (5.7e)$$

and the torsional moment for $\xi^{\bar{x}}{}_{\bar{y}} = z$

$${}^{x}dN^{\bar{x}} = \left(n^{\bar{x}}{}_{\bar{y}}\right)_{z} dy = \int_{c} z\,{}^{x}d\sigma_{y} = \int_{c} z\left(\left(\sigma_{y}{}^{\bar{x}}\right)_{z}(dydz)^{x}\right) =$$

$$= \left(\int_{c} \left(-z\left(\sigma_{y}{}^{\bar{x}}\right)_{z}\right)^{1} dz\right)_{z} dy \qquad (5.7f)$$

The composants of the interacting stresses developed on the interfaces $dA^{y}$ result from a similar to the above procedure as follows:

*I. Normal: (Fig. 5.4a)*

Axial force for $\xi_{y\bar{y}} = 1$

$${}^{y}dN_{y} = \left(n_{y\bar{x}}\right)^{z} dx = \int_{c} {}^{y}d\sigma_{y} = \left(\int_{c} -\left(\sigma_{y}{}^{\bar{y}}\right)_{e} dz\right)^{z} dx \qquad (5.8a)$$

Bending moment for $\xi^{\bar{x}}{}_{y} = z$

$$^y dN^{\bar{x}} = (n^{\bar{x}}{}_{\bar{x}})_e dx = \int_c z\,^y d\sigma_y = \left(\int_c \left(z\left(\sigma_y{}^{\bar{y}}\right)_e\right)^z dz\right)_e dx \qquad (5.8b)$$

*II. Lateral: (Fig.5.4b)*

Lateral shear force for $\xi_{z\bar{z}} = 1$

$$^y dN_z = (n_{z\bar{x}})^{\bar{y}} dx = \int_c^y {}^y d\sigma_z = \left(\int_c (\sigma_z{}^{\bar{y}})_x\, dz\right)^{\bar{y}} dx \qquad (5.8c)$$

Linear di-force for $\xi^1{}_{\bar{z}} = z$

$$^y dN^{\bar{1}} = \left(n^{\bar{1}}{}_{\bar{x}}\right)_x dx = \int_c z\,^y d\sigma_z = \left(\int_c (z(\sigma_z{}^{\bar{y}})_x)^y\, dz\right)_x dx \qquad (5.8d)$$

*III. Tangential: (Fig.5.4c)*

Tangential shear force for $\xi_{x\bar{x}} = 1$

$$^y dN_x = (n_{x\bar{x}})^1 dx = \int_c {}^y d\sigma_x = \left(\int_c (\sigma_x{}^{\bar{y}})_{\bar{z}} dz\right)^1 dx \qquad (5.8e)$$

Torsional moment for $\xi^y{}_{\bar{x}} = z$

$$^y dN^y = (n^y{}_{\bar{x}})_z dx = \int_c z\,^y d\sigma_x = \left(\int_c (z(\sigma_x{}^{\bar{y}})_{\bar{z}})^1\, dz\right)_z dx \qquad (5.8f)$$

## 5.6. Hidden Composants

As hidden are considered the composants of the hidden stresses resultants which are developed on the interfaces $dA^z = (dxdy)^z$ between the material points of the particle along the lateral axis $z$. The resultants of the hidden stresses are: (Fig. 5.6)

$$ {}^{z}\sigma_x = \left(\left(\left((\sigma_x{}^{\bar{z}})_{y,\bar{z}}\right)^{\bar{x}} dz\right)_y dA^z\right)_x \tag{5.9a} $$

$$ {}^{z}\sigma_y = \left(\left(\left((\sigma_y{}^{\bar{z}})_{\bar{x},\bar{z}}\right)^{\bar{y}} dz\right)_{\bar{x}} dA^z\right)_y \tag{5.9b} $$

$$ {}^{z}\sigma_z = \left(\left(\left((\sigma_z{}^{\bar{z}})_{e,\bar{z}}\right)^{z} dz\right)_e dA^z\right)_z \tag{5.9c} $$

The composants of the hidden stress resultants are represented by the general expression $dH_I = \left((h_I{}^{\bar{z}})dA^z\right)_I$ ,where $h_I{}^{\bar{z}}$ is their values: per unit area $A^z$. They are distributed in the following three classes and after taking into account the Eqs. (5.4), they are represented by the following particular expressions: (Fig. 5.6)

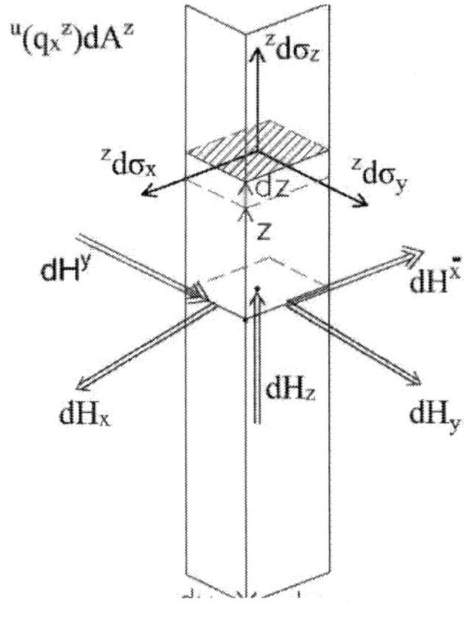


*Figure 5.6 - Hidden Composants*

I. *Polar vector hidden composants,* with $\xi_{I\bar{\imath}} = \xi_{x\bar{x}} = \xi_{y\bar{y}} = \xi_{z\bar{z}} = 1$

$$ dH_x = (h_x{}^{\bar{z}})dA^z = \int_c {}^{z}d\sigma_x = \int_c \left(((\sigma_x{}^{\bar{z}})_{y,\bar{z}})dz\right)_y dA^z = $$
$$ = \left({}^{u}\sigma_x{}^{\bar{z}} + {}_{\iota}\bar{\sigma}_x{}^{\bar{z}}\right)dA^z = (q_x{}^{\bar{z}})dA^z \tag{5.10a} $$

$$dH_y = (h_y{}^{\bar{z}})dA^z = \int_C {}^z d\sigma_y = \int_C \left(\left((\sigma_y{}^{\bar{z}})_{\bar{x},\bar{z}}\right) dz\right)_{\bar{x}} dA^z =$$
$$= \left({}^u\sigma_y{}^{\bar{z}} + {}_\iota\bar{\sigma}_y{}^{\bar{z}}\right)dA^z = (q_y{}^{\bar{z}})dA^z \tag{5.10b}$$

$$dH_z = (h_z{}^{\bar{z}})dA^z = \int_C {}^z d\sigma_z = \int_C \left(((\sigma_z{}^{\bar{z}})_{e,\bar{z}})dz\right)_e dA^z =$$
$$= \left({}^u\sigma_z{}^{\bar{z}} + {}_\iota\bar{\sigma}_z{}^{\bar{z}}\right)dA^z = (q_z{}^{\bar{z}})dA^z \tag{5.10c}$$

II. *Axial vector hidden composants*, with $\xi^I{}_{\bar{\iota}} = \xi^y{}_{\bar{x}} = \xi^{\bar{x}}{}_{\bar{y}} = z$

$$dH^y = ((h^{y\bar{z}})^x dA^z)^y$$
$$= \int_C \left(z({}^z d\sigma_x)\right)^y_x = \int_C \left(z\left(\left(((\sigma_x{}^{\bar{z}})_{y,\bar{z}})^{\bar{x}} dz\right)_y dA^z\right)_x\right)^y =$$
$$= \left(\int_C (\bar{\sigma}_z{}^{\bar{z}})_y dz\right)^x dA^z + \left(\int_C \left((-z(\sigma_x{}^{\bar{z}})_y)^x{}_{,\bar{z}}\right)_y dz\right)_x dA^z =$$
$$= \left(\int_C \left((\bar{\sigma}_x{}^{\bar{z}})_y dz\right)^x dA^z\right)^y + \left(\left(-\frac{c_z}{2}\left({}^u\sigma_x{}^{\bar{z}} + {}_\iota\sigma_x{}^{\bar{z}}\right)_y\right)^x dA^z\right)^y =$$
$$= \left(\int_C \left((\bar{\sigma}_x{}^{\bar{z}})_y dz\right)^x dA^z\right)^y + ((q^{y\bar{z}})^x dA^z)^y$$

or

$$h^{y\bar{z}} + \bar{q}^{y\bar{z}} + n_{z\bar{y}} = 0 \tag{5.10d}$$

From a similar procedure results

$$h^{\bar{x}\bar{z}} + \bar{q}^{\bar{x}\bar{z}} - n_{z\bar{x}} = 0 \tag{5.10e}$$

III. *Real scalar hidden composant,* with $\xi^{\bar{1}}{}_z = z$

From a procedure similar with that applied for the Eq. (5.10d) results:

$$dH^{\bar{1}} = \left(h^{\bar{1}\bar{z}}\right)^{z} dA^{z} = \int_{C} \left(z\left({}^{z}d\sigma_{z}\right)\right)^{1} = -\int_{C} \left(\left(\bar{\sigma}_{z}{}^{\bar{z}}\right)_{e} dz\right) dA^{z} + \left(q^{\bar{1}\bar{z}}\right)^{z} dA^{z}$$

or

$$h^{\bar{1}\bar{z}} - \int((\sigma_{z}{}^{\bar{z}})_{e} dz)^{z} = q^{\bar{1}\bar{z}} \tag{5.10f}$$

### 5.7. Inter-composant Relations

#### 5.7.1. Inter-tangential Composants Relations

From Eqs. (5.9a, b), (5.12a, b) and after taking into consideration the Cauchy equilibrium conditions (4.30), the following relations result:

$$n_{x\bar{x}} = n_{y\bar{y}} \tag{5.11a}$$

$$n^{y}{}_{\bar{x}} = n^{\bar{x}}{}_{\bar{y}} \tag{5.11b}$$

#### 5.7.2. Axial to Polar Vector Composants Relations

From Eqs. (5.2), (5.7b, c) and (5.4d, e), after taking into account the equilibrium conditions of stresses (4.29), and integration by parts the following differential equations between axial and vector composants are obtained:

$$\left(n^{y}{}_{\bar{x},\bar{y}}\right)^{x} = \int_{C} (((z(\sigma_{x}{}^{\bar{y}})_{\bar{z}})^{1} dz)_{z,\bar{y}})^{x} = \int_{C} \left(\left(-z((\sigma_{x}{}^{\bar{y}})_{\bar{z},\bar{y}})^{\bar{x}}\right)_{y} dz\right)^{x} \tag{5.12a}$$

$$\left(n^{y}{}_{\bar{y},\bar{x}}\right)^{\bar{x}} = \int_{C} (((-z(\sigma_x{}^{\bar{x}})_e)^{\bar{z}}dz)_{\bar{e},\bar{x}})^{\bar{x}} = \int_{C} \left(\left(-z((\sigma_x{}^{\bar{x}})_{e,\bar{x}})^{x}\right)_{\bar{y}}dz\right)^{\bar{x}} =$$

$$= \int \left(\left(z\left(\left(\bar{\sigma}_x{}^{\bar{y}}\right)_{\bar{z},\bar{y}} + \left(\bar{\sigma}_x{}^{\bar{z}}\right)_{y,\bar{z}} + \bar{g}_{x\bar{e}}\right)^{\bar{x}}\right)_{\bar{y}} dz\right)^{\bar{x}} =$$

$$= \int_{C} \left(\left(z\left(\left(\bar{\sigma}_x{}^{\bar{y}}\right)_{\bar{z},\bar{y}}\right)^{\bar{x}}\right)_{\bar{y}} + \left(\left(z(\bar{\sigma}_x{}^{\bar{z}})_{y}\right)^{\bar{x}}{}_{,z}\right)_{\bar{y}} - (\sigma_x{}^{\bar{z}})_y + (z(\bar{g}_{x\bar{e}})^{\bar{x}})_{\bar{y}}dz\right)^{\bar{x}} =$$

$$= \left(-\bar{n}^{y}{}_{\bar{x},y}\right)^{\bar{x}} + \left(\frac{C_z}{2}\left({}^{u}\bar{\sigma}_x{}^{\bar{z}} + {}_{l}\bar{\sigma}_x{}^{\bar{z}}\right)_y\right)^{\bar{x}}$$

$$+ \int_{C} ((\sigma_z{}^{\bar{x}})_{\bar{y}}dz)^{\bar{x}} - \bar{g}^{y\bar{z}} =$$

$$= -n^{y}{}_{\bar{x},\bar{y}} - \bar{q}^{y\bar{z}} - n_{z\bar{y}} - \bar{g}^{y\bar{z}} \tag{5.12b}$$

and finally

$$n^{y}{}_{\bar{x},\bar{y}} - n^{y}{}_{\bar{y},\bar{x}} + q^{y\bar{z}} + g^{y\bar{z}} = n_{z\bar{y}} \tag{5.12c}$$

From a similar procedure results

$$n^{\bar{x}}{}_{\bar{y},\bar{x}} - n^{\bar{x}}{}_{\bar{x},\bar{y}} + q^{\bar{x}\bar{z}} + g^{\bar{x}\bar{z}} = n_{z\bar{x}} \tag{5.13}$$

### 5.7.3. Polar Vector Composants to Loads Relations

From Eqs. (5.7a), (5.8e), after taking into account the Eqs. (5.2), (5.4), the equilibrium conditions of stresses (4.29), the following relations are obtained:

$$\left(n_{x\bar{y},\bar{x}}\right)_y = \int_{C} (((\sigma_x{}^{x})_e dz)^{\bar{z}}{}_{,\bar{x}})_y = \int_{C} \left(-((\sigma_x{}^{\bar{x}})_{e,\bar{x}})^{x}dz\right)_y =$$

$$= \int_{C} \left(\left(\left(\bar{\sigma}_x{}^{\bar{y}}\right)_{\bar{z},\bar{y}} + \left(\bar{\sigma}_x{}^{\bar{z}}\right)_{y,\bar{z}}\right)^{\bar{x}} dz\right)_y + \int_{C} ((\bar{g}_{x\bar{e}})^{\bar{x}}dz)_y \tag{5.14a}$$

$$\left(n_{x\bar{x},\bar{y}}\right)_{\bar{y}} = \int_c \left(((\sigma_x{}^{\bar{y}})_{\bar{z}} dz)^1{}_{,\bar{y}}\right)_{\bar{y}} = \int_c \left(-((\sigma_x{}^{\bar{y}})_{\bar{z},\bar{y}})^{\bar{x}} dz\right)_{\bar{y}} \qquad (5.14b)$$

and

$$\left(n_{x\bar{y},\bar{x}} - n_{x\bar{x},\bar{y}}\right)_y = \int_c \left(\left(\left(\bar{\sigma}_x{}^{\bar{z}}\right)_{y,\bar{z}}\right)^{\bar{x}} dz\right)_y = \int_c ((\bar{g}_{x\bar{e}})^{\bar{x}} dz)_y =$$

$$= \left({}^u\bar{\sigma}_x{}^{\bar{z}} + {}_\iota\bar{\sigma}_x{}^{\bar{z}}\right)_y + \left(\bar{g}_x{}^{\bar{z}}\right)_y = \left(\bar{q}_x{}^{\bar{z}} + \bar{g}_x{}^{\bar{z}}\right)_y$$

or

$$n_{x\bar{y},\bar{x}} - n_{x\bar{x},\bar{y}} + q_x{}^{\bar{z}} + q_x{}^{\bar{z}} = 0 \qquad (5.14c)$$

From a similar procedure applied on Eqs. (5.8a) and (5.7e) the corresponding relation between $n_{y\bar{x}}$ and $n_{y\bar{y}}$

$$n_{y\bar{x},\bar{y}} - n_{y\bar{y},\bar{x}} + q_y{}^{\bar{z}} + g_y{}^{\bar{z}} = 0 \qquad (5.15)$$

as well as, from Eqs. (5.78c) and (5.8c), the corresponding relation between $n_{z\bar{x}}$ and $n_{z\bar{y}}$,

$$n_{z\bar{x},\bar{y}} + n_{z\bar{y},\bar{x}} + q_z{}^{\bar{z}} + g_z{}^{\bar{z}} = 0 \qquad (5.16)$$

are obtained.

### 5.7.4. Real Scalar Interacting to Hidden Composants Relation

From Eqs. (5.7d), (5.8d), (5.10f), after taking into account the equilibrium (4.29) and integrating by parts, the following relations are obtained:

$$\left(n^{\bar{1}}{}_{\bar{y},\bar{x}}\right)^z = \int_c \left(\left((-z(\sigma_z{}^{\bar{x}})_{\bar{y}})^{\bar{x}} dz\right)_{y,\bar{x}}\right)^z =$$

$$= \int_c \left(\left(-z((\sigma_z{}^{\bar{x}})_{\bar{y},\bar{x}})^{\bar{z}}\right)_{\bar{e}} dz\right)^z \qquad (5.17a)$$

$$\left(n^{\bar{1}}{}_{\bar{x},\bar{y}}\right)^{\bar{z}} = \int_c \left(((z(\sigma_z{}^{\bar{y}})_x)^y dz)_{x,\bar{y}}\right)^{\bar{z}}$$

$$= \int_c \left(\left(z((\sigma_z{}^{\bar{y}})_{x,\bar{y}})^{\bar{z}}\right)_e dz\right)^{\bar{z}} \qquad (5.17b)$$

and

$$\left(n^{\bar{1}}{}_{\bar{y},\bar{x}} - n^{\bar{1}}{}_{\bar{x},\bar{y}}\right)^z = \int_c \left(\left(-z((\sigma_z{}^{\bar{x}})_{\bar{y},\bar{x}} + (\sigma_z{}^{\bar{y}})_{x,\bar{y}})^{\bar{z}}\right)_{\bar{e}} dz\right)^z =$$

$$= \int_c \left(\left(z((\sigma_z{}^{\bar{z}})_{e,\bar{z}})^z\right)_{\bar{e}} dz\right)^z + \int_c ((-z(\bar{g}_{z\bar{e}})^{\bar{z}})_{\bar{e}} dz)^z =$$

$$= \int_c \left(\left(\left(z(\bar{\sigma}_z{}^{\bar{z}})_e\right)^z{}_{,\bar{z}}\right)_{\bar{e}} dz\right)^z + \int_c ((-(\sigma_z{}^{\bar{z}})_e)_{\bar{e}} dz)^z + \bar{g}^{\bar{1}\bar{z}} =$$

$$= \left(\frac{C_z}{2}\left({}^u\bar{\sigma}_z{}^{\bar{z}} + {}_l\sigma_z{}^{\bar{z}}\right)_e\right)^z + \bar{h}^{\bar{1}\bar{z}} + q^{\bar{1}\bar{z}} + \bar{g}^{\bar{1}\bar{z}} = \bar{q}^{\bar{1}\bar{z}} + \bar{g}^{\bar{1}\bar{z}}$$

or

$$n^{\bar{1}}{}_{\bar{y},\bar{x}} - n^{\bar{1}}{}_{\bar{x},\bar{y}} + q^{\bar{1}\bar{z}} + g^{\bar{1}\bar{z}} = 0 \qquad (5.17)$$

### 5.8. Motives

The motives $U_K$, with the index $k$ taking the values $x, y, z, {}^{\bar{x},}\ {}^{y,}\ {}^{\bar{1}}$ are refereed to the middle point to the middle point of the lateral axis of the element in the $z$-direction (Fig 5.7.) and they correspond to the surface and inertial loads composants $\left((q_K{}^{\bar{z}})dA^z\right)_K$, $\left((g_K{}^{\bar{z}})dA^z\right)_K$ respectively, as well as to the hidden composants $\left((h_K{}^{\bar{z}})dA^z\right)_K$.

The motives at the interfaces of the particle, which correspond to the interacting composants, result from the motives $U_k$ in the following manner. At the two interfaces $A^x$ in the positions $x$ and $x + dx$ the motives corresponding to $U_K$ will be $U_K + \frac{\bar{U}_{K,x}dx}{2}$ and $U_K + \frac{U_{K,\bar{x}}dx}{2}$ respectively. In a similar manner the motives corresponding to $U_k$ at the interfaces $A^y$ in the positions $y$ and $y$+$dy$ will be $U_K + U_{K,y}dy$ and $U_K + U_{K,\bar{y}}dy$ respectively. In (Fig. 5.7) are shown the motives at the interfaces which correspond to the central motive $U_x$.

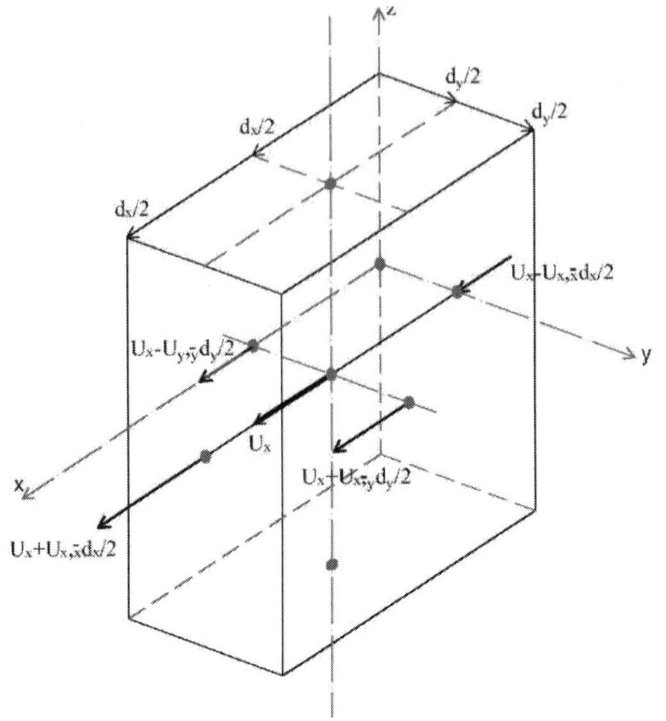


*Figure 5.7 - Motive $U_x$*

In cases where a distinction is required between the motives in the two directions the following notation will be used. The motives at the interfaces $A^x$ in the positions $x$ and $x+dx$ will be presented by the respective symbols

$$U_I = U_K + \overline{U}_{K,x}\, dx/2, \qquad U_I + U_{I,x}dx = U_K + U_{K,\bar{x}}dx/2 \qquad (5.18a)$$

While the motives at the interfaces $A^y$ in the positions $y$ and $y+dy$ will be presented by the symbols

$$U_J = U_K + \overline{U}_{K,\bar{y}}\, dy/2, \qquad U_J + U_{J,\bar{y}}dy = U_K + U_{K,\bar{y}}\, dy/2 \qquad (5.18b)$$

## 5.9. Universal Strains and Stresses

The variation of the work performed by the composants for the corresponding motives along the widths $dx$ and $dy$ of a particle is expressed by the following relations.

For the interacting composants ${}^{x}dN_I = (n_{I\bar{y}})dy$ developed on the interfaces $dA^x = (dy)c_z$ and the motives $U_I$

$$\Delta_I^N W^{\bar{1}} = \left((n_{I\bar{y}})dy\right)_I [x+dx] \cdot U_I[x+dx] - \left((n_{I\bar{y}})dy\right)_I [x] \cdot U_I[x] =$$
$$= \left((n_{I,\bar{y}})dy + \left((n_{I\bar{y}})dy\right)_{,\bar{x}} dx\right)_I \cdot (U_I + U_{I,\bar{x}}dx) - \left((n_{I\bar{y}})dy\right)_I \cdot U_I =$$
$$= \left((n_{I\bar{y}})dy\right)_I \cdot (U_{I,\bar{x}}dx) + \theta_{(Iy)x} U_I \cdot \left(\left((n_{I\bar{y},\bar{x}})dx\right) dy\right) \quad (5.19a)$$

and in a similar manner, for the interacting composants ${}^{y}dN_J = (n_{J\bar{x}})dx$ developed on the interfaces $dA^y = c_z(dx)$ and the motives $U_J$

$$\Delta_J^N W^{\bar{1}} = \left((n_{J\bar{x}})dx\right)_J \cdot (U_{J,\bar{y}}dy)_J + \theta_{(Jx)y} U_J. \left(\left((n_{J\bar{x},\bar{y}})dx\right) dy\right) \quad (5.19b)$$

The work performed by the hidden composants $dH_K = (h_K{}^{\bar{z}})(dxdy)^z$ for the motives $U_K$ is

$$\Delta_J^H W^{\bar{1}} = \left((h_K{}^{\bar{z}})(dxdy)^z\right)_K \cdot U_K \quad (5.19c)$$

and the work performed by the external and the inertial loads

$$\Delta^{(q+g)}{}_K W^{\bar{1}} = \left((q_K{}^{\bar{z}} + g_K{}^{\bar{z}})(dxdy)^z\right)_K \cdot U_K \quad (5.19d)$$

The variations of the work performed by the interacting and the hidden composants have to be equal with that performed by the external and inertial loads, namely

$$\Delta_I^N W^{\bar{1}} + \Delta_J^N W^{\bar{1}} + \Delta_K^H W^{\bar{1}} = \Delta^{(q+g)}{}_K W^{\bar{1}} \quad (5.20a)$$

After replacing Eqs. (5.19a to d) to Eq. (5.20a) and taking into account the inter-composant relations, Eq. (5.20a) takes the expression

$$\frac{1}{2}\left(\left((n_{I\bar{y}})dy\right)_I \cdot (U_{I,\bar{x}}dx) + \left((n_{J\bar{x}})dx\right) \cdot (U_{J,\bar{y}}dy)\right) =$$
$$= \left((q_K{}^{\bar{z}} + g_K{}^{\bar{z}})(dxdy)^z\right)_K \cdot U_k \qquad (5.20b)$$

From Eq. (5.20b) results that to the work performed by the external and inertial loads corresponds the part of the work performed by the interacting composants which is represented by the first terms of the right side of Eqs. (5.19a, b). The factor ½ is due to the fact that the composants $n_{I\bar{y}}$ and $n_{J\bar{x}}$ are developed simultaneously with the deformations $U_{I,\bar{x}}$ and $U_{J,\bar{y}}$ respectively under a relation considered as linear along the infinitecimal widths $dx$ and $dy$.

The omitted second terms at the right side of Eqs. (5.19a, b), which are equal and opposite to the right side of Eq. (5.19c), represent energy quantities exchanged between the interacting and the hidden composants.

The entities $U_{I,\bar{x}}$ and $U_{J,\bar{y}}$ represent the rate of variation of the motives $U_I$ and $U_J$ along the axes $x$ and $y$ respectively and will be referred as *universal strains*. For the different values of the indices $I$ and $J$ they are defined in Structural Mechanics by the following terms and notations:

| | |
|---|---|
| $U_{x,\bar{x}}$ and $U_{y,\bar{y}}$ : | Mean normal strains $\varepsilon_{xxo}$ and $\varepsilon_{yyo}$. |
| $U_{z,\bar{x}}$ and $U_{z,\bar{y}}$ : | Mean lateral shear strains $\varepsilon_{zxo}$ and $\varepsilon_{zyo}$. |
| $U_{x,\bar{y}}$ and $U_{y,\bar{x}}$ : | Mean tangential shear strains $\varepsilon_{xyo}$ and $\varepsilon_{yxo}$. |
| $U^{\bar{x}}{}_{\bar{y}}$ and $U^{y}{}_{,\bar{x}}$ : | Flexural curvatures $\kappa_x$ and $\kappa_y$. |
| $U^{x}{}_{\bar{\bar{x}}}$ and $U^{y}{}_{,\bar{y}}$ : | Torsional curvatures. |
| $U^{\bar{1}}{}_{\bar{x}}$ and $U^{1}{}_{,y}$ : | NA. |

The entities $n_{I\bar{y},\bar{x}}$ and $n_{J\bar{x},\bar{y}}$ represent the rate of variation of the composants $n_{I\bar{y}}$ and $n_{J\bar{x}}$ along the axes $x$ and $y$ respectively and will be referred as *universal stresses*. A definition of them is not represent in the present state of Structural Mechanics because the hidden composants and the associated with them hidden (or dark) exchange energy are ignored.

# 6. THREE-DIMENSIONAL AGGREGATES OF PARTICLES

## 6.1 Zero-Dimensional Particles

In the previous chapters 4 and 5 the cases of one dimensional aggregates of two dimensional particles and of two dimensional aggregates of one dimensional particles respectively are discussed. This sequence is normally completed by the consideration of three dimensional aggregates of zero dimensional particles occupying infinitesimal volumes $dV^e = (dxdy)dz$. On these particles the inertial forces $dg_i = g_i^{\bar{e}} dV^e$ and the interacting ones

$$d\sigma_i = \sum_j \sigma_i^{\bar{j}} dA^j$$

as well as the corresponding momentums $dp_i = p_i^{\bar{e}} dV^e$ and interacting impulses

$$d\tau_i = \sum_j \tau_i^{\bar{j}} dA^j$$

are applied, executing translations and orbital velocities respectively, all of them represented by polar vectors. Forces, momentums, displacements and velocities represented by scalar or axial vectors result from the polar vector entities multiplied by infinitesimal lever-arm lengths included into the infinitesimal volume and constitute higher order infinitesimals which are omitted. Under these conditions a three dimensional aggregate of zero dimensional particles with full contact connections between them coincides with the deformable continuum with particles the material points already discussed in chapter 2.

## 6.2 Solid Particles of Finite Size

As non-zero dimensional solid particles constituting a composite three dimensional aggregate can be considered the simple three dimensional particles as well as the one dimensional aggregates of two dimensional particles and the two dimensional aggregates of one dimensional particles presented in the previous chapters 3,4 and 5 respectively. The connections between the constituent parts is not necessarily realized by a full contact over their total boundary surfaces. They are usually restricted to local contact surfaces with connections appearing various degrees of freedom concerning the relative motion between the connected parts, which are expressed by the constrain conditions between their motives at the interfaces. Rigid jointed, pin jointed or sliding connections are represented by the constrain conditions, which define the restrained DOF of the relative motion in the three principal directions of the contact surface. The natural granular and block aggregates as well as the technological structures consisting of bars, beams, plates and members of machines are contained to this category.

A connection is considered between two particles with its location defined by the two position vectors $r$ of the center of the contact surface in the two local reference systems of the particles. The total composant of the interacting stresses developed on the contact surface is similar to the total interacting composant of the two dimensional particles and will be furthermore referred as *reaction* having the general form

$$R = R_p + R^a + R_1 + R^e \qquad (6.1)$$

The two total interacting composants $F = F_p + F^a + F_1 + F^e$ due to the reaction $R$ each one with respect to the center of mass of the two connected particles, are estimated after considering the reaction $R$ equal to a superposition of two sets of forces applied at the points $r$ and $-r$ of each one particle, the one symmetric corresponding to the parts ${}_sF$ of $F$ and the other antisymmetric corresponding to the part ${}_aF$ of $F$, and (Fig 6.1.a to d). In the symmetric set the forces $R/2$ are applied at both points $r$ and $-r$ while in the antisymmetric set the force $R/2$ is applied at the point $r$ and the force $-R/2$ at the point $-r$. From (Fig. 6.1) becomes apparent that ${}_sF + {}_aF = R$.

The part of ${}_aF$ due to the vector components $R_p$ and $R^a$ of $R$ results after analyzing each one of them in a radial component $R_r = \hat{r}(r \cdot R_p)_1$, $R^r = r(r \circ R^a)^e$ and in a transversal one $R_t = \hat{r}(r \times R_p)$, $R^t = \hat{r}(r \otimes R^a)$ lying on the planes $(r, R_p)$ and $(r, R^a)$ respectively (Figs. 6.1a, b). The radial components correspond to the scalar composants

$${}_aF_1 = rR_r = (r \cdot R_p)_1, \qquad {}_aF^e = rR^r = (r \circ R^a)^e \qquad (6.2a)$$

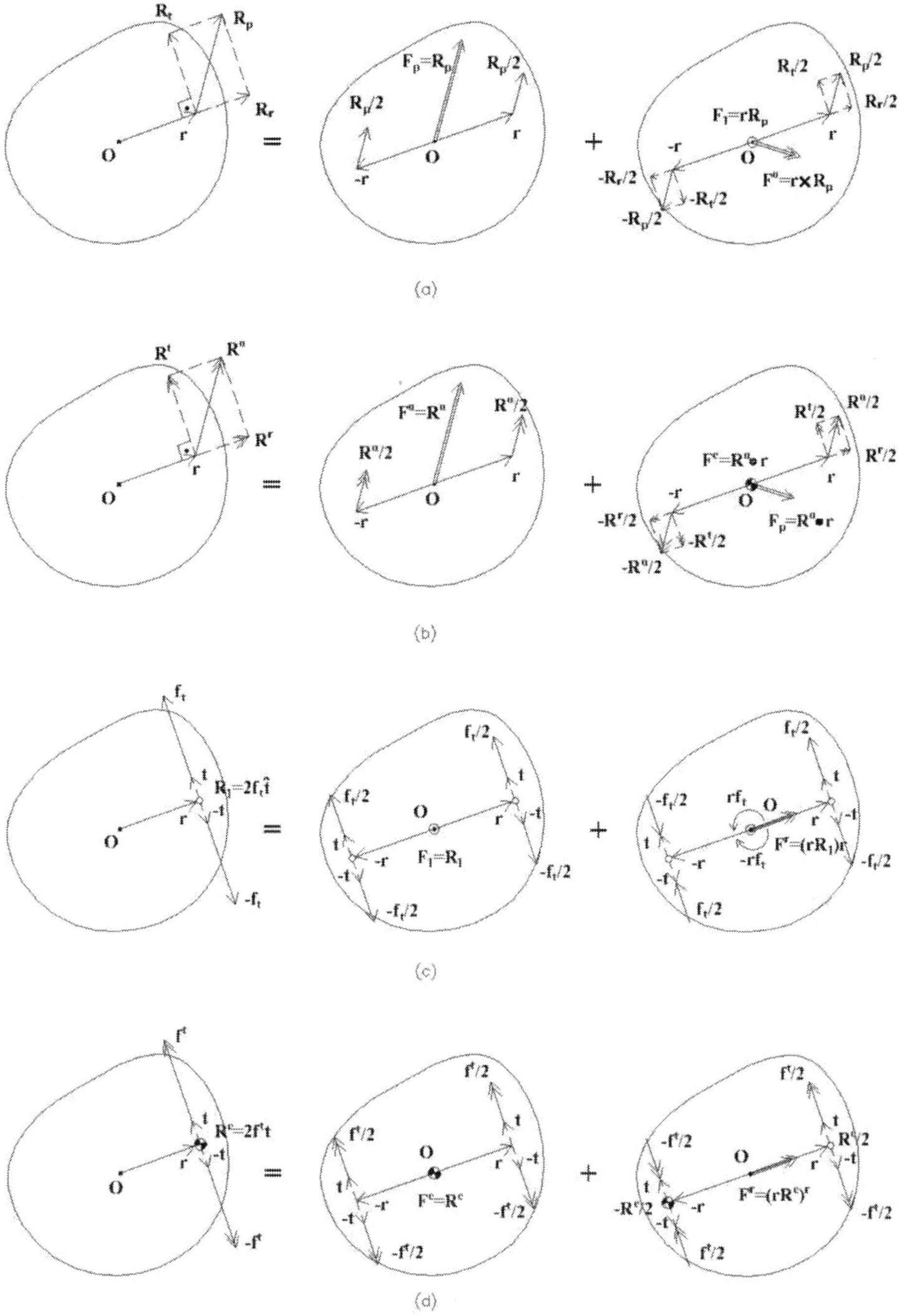


*Figure 6.1 - Octonionic Composants*

while the transversal components correspond to the vector composants

$$ {}_aF^q = rR_t = \left(r \times R_p\right)^q, \qquad {}_aF_{q'} = rR^{t'} = (r \otimes R^a)_{q'} \tag{6.2b} $$

where $e^q = e_r e_t$ and $e_{q'} = e_r e^{t'}$.

The part of ${}_aF$ due to the scalar components $R_1$ and $R^e$ of $R$ results after replacing each one of them by equivalent pairs of forces ${}^{\pm}f_t$ acting in directions transversal to $r$ at the points $r \pm t$ respectively, so that $R_1 = 2f_t\hat{t}$ and $R^e = 2f^t t$. (Figs. 6.1c, d). To the antisymmetric set of $f_t$ corresponds the contra-formal polar vector composant

$$ {}_aF_{p2} = 2(rf_t)\hat{t} = 2r(f_t\hat{t})_1 = (rR_1)_r \tag{6.2c} $$

and to the antisymmetric set of $f^t$ corresponds the axial vector composant

$$ {}_aF^{a2} = (rR^e)^r \tag{6.2d} $$

Finally the total interacting composant $F$ will be

$$ F = {}_sF + {}_aF = R + {}_aF_1 + {}_aF^e + {}_aF^q + {}_aF_{q'} + {}_aF_r + {}_aF^r = $$
$$ = R + r \cdot R_p + r \circ R^a + r \times R_p + r \otimes R^a + rR_1 + rR^e $$

or, in terms of the octonion product, the part ${}_iF$ of the total composant $F$, due to the raction ${}_ir$, will be

$$ {}_iF = {}_iR + {}_ir\,{}_iR = \left(1 + {}_ir\right){}_iR \tag{6.3} $$

If ${}_oF$ is the total composant of the forces acting on the particle considered, its total composant will be

$$ F = {}_oF + \sum_i {}_iF = {}_oF + \sum_i \left(1 + {}_ir\right){}_iR \tag{6.4} $$

The work executed by the total composant $F$ for the corresponding to it motive $U$ is

$$W_{\bar{1}} = U \cdot F = U \cdot {}_oF + U \cdot \sum_i (1 + {}_ir)\ {}_iR$$
$$= U \cdot {}_oF + \sum_i U(1 + {}_ir) \cdot {}_iR =$$
$$= U \cdot {}_oF + \sum_{\Sigma} {}_i\Upsilon \cdot {}_iR \qquad (6.5)$$

where

$${}_i\Upsilon = U(1 + {}_ir) \qquad (6.6)$$

is the motive of the contact surface at the point ${}_ir$, where the reaction ${}_iR$ is applied

### 6.3 Solid Particles of Infinite Size

The totality of the material points of an infinite deformable solid continuum is considered as a solid particle of infinite size, in case the points execute translations

$$u_p = \sum_i u_i$$

due to deformation, with values defined by the relations

$$u_i = \sum_I U_I \zeta_{\bar{I}i}$$

where $\zeta_{\bar{I}i}$ are shape functions and $U_I$ the elements of a total octonion motive

$$U = \sum_I U_I = U_1 + \sum_i U_i + \sum_i U^i + U^e = U_1 + U_p + U^a + U^e$$

which is referred to a specific point of the continuum. Regions of the continuum bounded by a closed surface can be considered as quasi infinite in case the translations on the boundary surface appear negligible values.

A total composant

$$F = F_1 + F_p + F^a + F^e \sum_I \sum_i \xi_{I\bar{i}} df_i$$

Corresponds to the motive $U$, applied on the same reference point, which represents the totality of forces

$$df_i = \left( \sum_j \sigma_i{}^{\bar{J}}{}_{,\bar{J}} + g_i{}^{\bar{e}} \right) dV^e$$

distributed in the material points of infinitesimal volume $dV^e$, equal to the resultants of field forces $g_i^e dV^e$ and of the variations $\sigma_i{}^{\bar{J}}{}_{,\bar{J}} dV^e$ of the stresses developed on their sides.

Systems of translations and forces represented by different motives and composants, which are referred to the same or to different points, have to be considered as different particles, although they are applied on the same material points of the deformable solid continuum. Their aggregate consists to the superposition of the translations and forces applied on the totality of the material points and the distinction of separate regions of the continuum, each one corresponding t a distinct particle, is not furthermore generally possible.

Nevertheless for the usual cases, where the modules $|u_p|$ decrease with the distance $|r|$ from the reference points with the minimum values tending to zero at the infinite, a distinction of *domains* associated with the distinct particles can be established, which concern regions of the continuum where the values of the superposed translations are dominated from those of the dominant particle. The domains are separated by surfaces containing the points where the total translations receives minimum values and are defined by the equation

$$|u_p[r]|_{\bar{r}} = 0 \qquad (6.7)$$

About the author

Dimitris Bairaktaris was born in Athens in 1937. He received the first place award in the national competition amongst high school graduates from the Hellenic Mathematical Society. In 1960 he received his graduate degree in Civil Engineering (MSc) from the National Technical University of Athens and in 1969 he received his PhD in Structural Mechanics. From 1960 to 1975 he worked as an assistant professor and from 1975 to 1981 he was an associate professor at the same university. During that time he worked independently as a structural designer and continued his personal development in higher and modern mathematics and theoretical physics. In 1980 he became the founder and senior partner of “D. Bairaktaris and Associates Ltd – Structural and Geotechnical Consultants”, a company that has overseen the design of numerous complex buildings, bridges, and underground facilities both in Greece and abroad. Throughout the years he has also participated in a number of research projects, funded by the EU. As a research partner, he has been responsible for the development of the methodology and software for the structural condition assessment of buildings, bridges and tunnels, following natural or man-made destructive events. These were based on data acquired from measurements from suggested agglomerates monitoring as well as robotic inspection, also overseen by Dr. Bairaktaris’ team. He has presented over 100 publications, conference papers and research reports with technical, mathematical and academic content. To this day, he remains vigilant, active and productive in his field with an uncompromised passion for mathematics and theoretical physics that has guided his work on this book and others to come.

Printed in Dunstable, United Kingdom

91349757R10118